J. L Robinson

A Treatise on Marine Surveying

Prepared for the Use of Younger Naval Officers

J. L Robinson

A Treatise on Marine Surveying
Prepared for the Use of Younger Naval Officers

ISBN/EAN: 9783337156084

Printed in Europe, USA, Canada, Australia, Japan

Cover: Foto ©ninafisch / pixelio.de

More available books at **www.hansebooks.com**

A TREATISE

ON

MARINE SURVEYING:

PREPARED FOR THE USE OF YOUNGER NAVAL OFFICERS,

WITH QUESTIONS FOR EXAMINATION AND EXERCISES,
PRINCIPALLY FROM THE PAPERS OF THE
ROYAL NAVAL COLLEGE. WITH THE RESULTS.

REV. J. L. ROBINSON, B.A.

ROYAL NAVAL COLLEGE.

"There is, I am persuaded, no one among us .... who thinks so highly of himself, as not to believe that he may learn much, and derive much assistance from communication with his brethren, nor so engrossed with his own share of the common work, as not to be desirous of imparting to others whatever has been recommended by his own experience to himself."—THIRLWALL.

London:
MACMILLAN AND CO.
1882

[The right of Translation and Reproduction is reserved.]

Cambridge:
PRINTED BY C. J. CLAY, M.A. & SON,
AT THE UNIVERSITY PRESS.
2253

PREFACE.

THE following Treatise on Marine Surveying and its connected subjects is intended to meet an Educational want which has existed for some time past. I mean a small Manual on the subject, arranged on a plan similar to that employed with success in Text-books of Elementary Science, and which combining a sufficient description of the Instruments used, and of the various methods pursued, with a careful selection of representative Examples, *with their solutions*, may, it is hoped, serve as a sufficient Introduction to this branch of Science for the use of Younger Naval Officers, and also, if it may be, to bring the subject, at least in its more popular features, before a wider circle than those professionally interested in it.

A glance at the way in which the various subjects are treated, will shew that I have had no intention to write—or, if the word be preferred, to compile—a Hand-book for the use of the Practical Surveyor. Such an intention might fairly be regarded as an impertinence in one who has never been engaged in the practical work of the Profession, and who must therefore be ignorant of many details which of necessity enter largely into any book intended to serve that purpose.

I have rather had the Examination-room and its requirements before me in selecting and arranging the materials of my little book. It is well known that the scope of the

Examination in this subject has been greatly increased since the establishment of the Royal Naval College at Greenwich. Formerly, the only test proposed (irrespective of a *vivâ voce* Examination) was the construction of a very simple Mercator's Chart (vide a typical example, p. 91, No. 2), whereas now in addition to the construction of a chart, involving at times some peculiar difficulties (vide Nos. 11, 14, 16, 18, 22, in Chap. V.), questions are nearly always set which require a sound knowledge of the different methods of Fixing Positions, of the Management of Chronometers, of Meridian Distances, of the Use of Surveying Instruments, and of the varied and important information which is conveyed succinctly but with perfect accuracy by the Symbols employed in the Admiralty charts. It seemed to me that, under these circumstances, a small Manual embodying the necessary information would be welcomed by young officers who are called on to undergo an Examination in the subject. And if, in addition to meeting the wants of younger students, my other hope should be realized that the Book may succeed in enlisting the sympathy of readers for a most interesting and useful branch of study, in exciting a desire in Junior Naval Officers to gain, as opportunity offers, a knowledge of an important branch of their profession, and in laying a firm foundation for the future instruction, theoretical and practical, imparted in the subject in the Royal Naval College and elsewhere, then indeed I shall be rewarded for the many weeks and months of labour which its preparation has required.

And in this preparation I can conscientiously affirm that no pains have been spared to make it serve its original purpose of a Text-book for Young Students. The very best sources of information lay within my reach and have been freely drawn upon: whether these advantages have been utilized in the best way is, of course, a different question. Many kind friends have greatly assisted me by their judgment and technical knowledge. While I feel under obligations to others, I cannot sufficiently say how much I am indebted to Mr Oborn, the Lecturer in Navigation and Nautical Astronomy in the Royal Naval College. In addition to reading through the entire work in manuscript, he carefully corrected the proofs of the first eleven chapters. Many of his suggestions are embodied in the text, and I beg to thank him most sincerely for the great trouble that he has taken to make the little book accurate in its information. Staff-Commander Johnson, the Instructor in Nautical Surveying, in the Royal Naval College, gave me many valuable hints for the improvement of the Chapters on Instruments, Tides, Soundings, and Fixing Positions. These I gladly used either to correct or to supplement my own statements. And, finally, a friend of great experience as a First-class Surveyor read through and corrected several of the more important chapters. The assistance thus given so cheerfully, while of course not removing responsibility from me, cannot but cause me to entertain a greater confidence in submitting my little Book to the public than I should otherwise have been justified in feeling.

It only remains for me to add, that all possible care has been taken to insure accuracy in the Results appended to the Exercises. I shall, however, be very grateful to have my attention called to errors of any kind which may be found to exist, by those who use the book, as well as for any suggestions from Instructors, Students, or Practical Surveyors, which would, in their opinion, tend to make it more generally useful.

JOHN LOVELL ROBINSON.

ROYAL NAVAL COLLEGE,
July 21, 1882.

TABLE OF CONTENTS.

CHAPTER I.

SYMBOLS USED IN CHARTS, &c.

I. Symbols to denote the quality of the Bottom—Examination—II. General Abbreviations—Examination—III. Buoys—Colouring—Regulations for employing—Examination—IV. Lights and Lighthouses—Explanations and Illustrations—Illuminating Apparatus—Examination—V. Conventional Signs—Examination—VI. Soundings—VII. Miscellaneous.—Examination—VIII. Surveying Symbols—Authorities on this subject.
pp. 1—19

CHAPTER II.

CONSTRUCTION AND USE OF SCALES.

Definition—I. *Simply-divided Scales:* (α) Scales of Equal Parts—Use—Mode of constructing—Exercises—The Natural Scale—(β) Diagonal Scales—When employed—Exercises in constructing—General Exercises in Construction—Plain and Diagonal Scales—(γ) Verniers—General principles—Exercises in constructing; Principle of Reading off—To ascertain whether the Limb is correctly graduated—Mode of correcting.—II. *Protracting Scales:* Names and Uses of these—Construction and Use of a Scale of Chords—Scale of Rhumbs; The Sector—Sectoral Lines—Use of these—Marquois Scales—Use—Exercises—To erect a Perpendicular—To divide a Line into a number of Equal parts: (i) By Trial—(ii) By Proportional Compasses—(iii) By a Scale of Equal parts—(iv) By Marquois Scales . . . . pp. 20—48

x   CONTENTS.

CHAPTER III.
LAYING OFF ANGLES.
Methods of Protracting Angles: (1) By Semicircular Protractor. (2) By Rectangular Protractor. (3) By Circular Protractor. (4) By Scale of Chords. (5) By Computing the Length of a Chord to a given Radius. Proof of the Rule employed—Examples for Exercise . . pp. 49—52

CHAPTER IV.
FIXING POSITIONS.
Necessary Geometrical Theorems—To describe a segment capable of containing any given angle: (i) When the Angle is right; (ii) When it is acute; (iii) When it is obtuse—Exercises—Two methods of fixing a position: (α) The straight line and one angle method. (β) The Three-point Problem—*Six different cases* of the Three-point Problem: (i) When P is outside the triangle formed by joining the three points A, B, C, and the central object is on the side of the line joining the other two points, remote from P—Two methods—Notes on the second method—(ii) When P lies outside the triangle, but the central object and P lie on the same side of the line joining the other two objects. Note—(iii) When P is on one side produced of the triangle formed by joining the three objects A, B and C—(iv) When P is on one side of the triangle ABC—(v) When the three objects A, B, C are in the same straight line—Two methods—(vi) When P is inside the triangle ABC—Two methods—The Indeterminate Case ("on the Circle")—Note on this Case—Best relative position of the objects when using the Three-point Problem—Fixing Positions without describing segments—Lines of Position—Straight-line method—Exercises—Danger Angle—Useful Problems: (i) To find the length of a line accessible only at the ends—(ii) To find the distance of an inaccessible object by the rhombus method—(iii) To find the distance of the point where two lines intersect in a river or lake—Examination—Exercises—Answers . . . . . pp. 53—80

CHAPTER V.
CHARTS AND CHART DRAWING.
Definitions—Projection of a point—Orthographic Projection—Stereographic Projection—Central or Gnomonic Projection—Mercator's Projection—Plane Chart—Nautical Mile—Advantage and Disadvantage of Mercator's Chart—Method of Constructing a Mercator's Chart—Erection of the Perpendiculars—To lay down the Latitude and Longitude of a Point on a Chart—To lay down a Course on a Chart—To lay down a Distance on a Chart—Two special points noticed more particularly—Exercises—Answers . . . . pp. 81—100

CONTENTS. xi

CHAPTER VI.

INSTRUMENTS AND OBSERVING.

Gunter's Chain—Ordinary Surveying Chain—Sextant—Adjustments—To test the Adjustments—Method of correcting error in Collimation—Precaution in observing—Theodolite—Three principal parts—Three principal motions—Three Adjustments—Method of using—To take an "Arc"—To repeat an Angle—To observe a Vertical Angle—The Spirit Level—Description—Three Adjustments—The Levelling Staff—The Ten-feet Pole—Method of Computing the Scale—The Station-Pointer—Method of using—The Barometer—Three Methods of correcting the height of the column: (i) By Capacity Correction—(ii) By Flexible Base—(iii) By Contracted Scale—The Marine Barometer—Method of Suspending—Vernier of the Marine Barometer—Method of reading off—Method of stowing for carriage . . . . pp. 101—120

CHAPTER VII.

BASE LINES.

Use of a Base—Base of Verification—Example—Geodetic Standards of Measurement vary—Standards "à traits" and "à bouts"—The Knowledge of the exact Length of a Bar at any moment involves three distinct elements. The temperature difficulty is evaded by Three Methods: (i) Borda's—(ii) Colby's—(iii) Struve's—Description of Colby's "Compensation Apparatus"—Note on the U. S. Coast Survey Apparatus—Ratio of length of Base to Area of Survey—Degree of accuracy required—The Base Lines in the Ordnance Survey—The Madrid Base—Reduction of Base to the Sea Level—The Three Elements of the Base—Precautions to be observed in selecting the ground—Different Methods of Measuring a Base: (i) Masthead Angle—(ii) Velocity of Sound—Objections to this Method—(iii) Patent Log under Steam—(iv) Astronomical Observations—(v) Direct Measurement—Method of proceeding—Formulæ—The Direction of the Base—Examination—Exercises—Answers . . . . pp. 121—139

CHAPTER VIII.

TRIANGULATION.

Object sought—Personal Equation—Method of "piling-up" the Triangles—Primary and Secondary Triangles—Observing Distant Points—Assumed Base—Spherical Excess—Legendre's Theorem—Reduction to the Centre—Triangulation of England connected with that of the Continent—Triangulation of Spain connected with that of Algiers—Examination . . . . pp. 140—148

CHAPTER IX.

LEVELLING.

Definitions—Check Level—Simple and Compound Levelling—To find difference of Level of two points—Method of running a line of Level through a tract of country—To lay down a Section—Examples—Levelling by Theodolite—By Barometer—Use of the Mountain Barometer—Hutton's Rule to compute the difference of Height by heights of the Barometer at the two stations—By Thermometer—Contours—Method of delineating—Examination . . . . pp. 149—165

CHAPTER X.

TIDES AND TIDAL OBSERVATIONS.

Theories of Bernouilli and Laplace—Tidal phenomena—Spring and Neap Tides—Superior and Inferior Tides—Cotidal Lines—Head or End of the Tide—Difference between Tidal Current and Tidal Stream—Flow and Ebb—Tide and Half Tide—Tide and Quarter Tide—Tide and Half Quarter Tide—Conjunction—Opposition—Quadrature—Lunation—Semi-lunation—Perihelion—Aphelion—Perigee—Apogee—Circumstances under which the Tides are greatest and least—Diurnal inequality—Effects of Wind and Barometer on the Tides—Rise and Range of a Tide—Mean Level of the Sea—Tide Day—Priming and Lagging of the Tide—Single Day Tides—Double Half Day Tides—Octant Tides—Age of the Tide—Retard—Lunitidal Interval—Vulgar Establishment of the Port—Corrected Establishment of the Port—Semimenstrual Inequality—Time and Height of High Water—Method of Interpolation—First and Second High Water—Tidal Constants—Methods of Computing the Time of High Water—Tidal Observations—Elements of a Complete Tidal Table—Tide Gauges—Precautions to be observed—Tidal Evolution—Examination . . . . . pp. 166—199

CHAPTER XI.

SOUNDINGS.

Objects sought—Sounding Pole—Fixing Boat at Starting—To keep the Boat on a certain "range"—Quality of the bottom to be registered. To fix a sounding—Lead Lines must be tested—Reduction of Soundings—Method of effecting this reduction—Formula for reducing—Soundings in a Tidal River—Elevations along the course of a Tidal River—To discover and fix the position of a hidden Danger—Surface Currents—Current Log—Under-Currents—Discharge of a River—Definitions—To find the Sectional Area—To find the Mean Velocity of the Current: (i) By Floats—(ii) By the Tachometer—Ground Log—Buoy and Nipper—Examination. . . pp. 200—211

CONTENTS. xiii

CHAPTER XII.

CHRONOMETERS.

Definitions—History—Harrison's Watches—The Inside of a Chronometer—Bi-metal Balance—Methods of adjusting—Secondary Error—Airy's Method of Compensation—Chronometer Room in the Royal Observatory—"Annual Trials"—Report by the Astronomer Royal—"Trial Numbers"—Chronometers for the Royal Navy—History of a Chronometer while in the Service—Packing a Chronometer for transport—To take a *Going* Chronometer from the Shore to a Ship—Two sources of error to be guarded against—Stowage of Chronometers on board—Effects of Temperature—Three reasons why a Chronometer should be wound at the same hour daily—Method of Winding a Chronometer—Comparison of Chronometers—Rates of Chronometer liable to change—To detect a faulty Chronometer when Three are in use—Exercises in comparing Chronometers—To compare a Chronometer with a Sidereal Clock—Examination . . pp. 212—234

CHAPTER XIII.

MERIDIAN DISTANCES.

Definition—Primary and Secondary Meridians—Two Methods of Computing the Meridian Distance: (i) Electric Telegraph—(ii) Portable Chronometers—Precautions to be observed in using the Second Method—To find P.M. time of Observations for Equal Altitudes—Examination—Formulæ for Sea Rate—Typical Examples worked out as Models—General Examination, and Examples for Exercise . . . . pp. 235—250

CHAPTER XIV.

METHOD OF PLOTTING.

Setting about the Work—Projecting the Work from the Field Book. Points to be noted in conducting a Survey—Examples of Plotting—General Exercises . . . . pp. 251—289

Miscellaneous Exercises . . . . p. 289

General Index . . . . p. 301

BOOKS CONSULTED.

The following Works have been carefully consulted in preparing the Materials of the following Treatise.

Admiralty Tide Tables, for the years 1880, 81, 82.
Admiralty List of Lights on the British Coasts, 1880, 81, 82.
Airy's "Tides and Waves" in the Encyclopædia Metropolitana.
Barometer Manual.
Belcher's Nautical Surveying.
Brinkley's Elements of Astronomy, edited by Stubbs and Brünnow.
Burr's Instructions in Practical Surveying.
Clarke's Geodesy.

This is a most valuable treatise for the more advanced Mathematical Student. It contains an account of the most recent methods pursued in important Trigonometrical Surveys.

Compendium of Instructions for Hydrographic Surveyors, 1877.
De Rheims. Geometrical Drawing, 1865.
Dubois. Cours de Navigation et d'Hydrographie.

This contains an excellent Chapter on Chronometers and their management.

Frome's Outline of the Method of conducting a Trigonometrical Survey.
Galbraith's (W.), Trigonometrical Surveying.
Galbraith and Haughton. Tides and Tidal Currents.
Harbord's Glossary of Navigation.
Harrison. An Account of the Goings of Mr J. Harrison's Watch, 1767.

Several very interesting works on this subject, of about the same date, may be seen in the British Museum.

Haskoll's Land and Marine Surveying.
Heather's Treatise on Mathematical and Drawing Instruments.

The original work formed one small volume in Weale's Series. It has lately been enlarged, and is now published in three volumes, containing respectively: "Drawing and Measuring Instruments;" "Optical Instruments;" "Surveying and Astronomical Instruments."

Herschell's Treatise on Astronomy.
Hull's Nautical Surveying.
Jeffers' Nautical Surveying.
Johnson's V. Staff-Com. Notes on Marine Surveying.
Laughton's Introduction to Nautical Surveying.
Ledieu. Les Nouvelles Méthodes de Navigation. Paris, 1877.

The Second Part of this work is devoted exclusively to the Description, Theory, and Use of Chronometers, and ought to be read by all who are interested in Chronometrical Science.

Manual of Scientific Inquiry. The Articles on Hydrography and the Tides.
Manual of Surveying for India.
Pearson's Elementary Treatise on Tides, 1881.
Philosophical Transactions for the period 1833—1847.

In these volumes are contained the Original Researches of Dr Whewell and Sir John Lubbock on the Tides. The volumes for 1879 and 1880 contain Professor Darwin's Theory of Tidal Evolution.

Raper's Practice of Navigation. 3rd Edition.
Shadwell's Notes on the Management of Chronometers.

I beg to acknowledge here my great obligation to this valuable and well-known Treatise. The useful information embodied in my own Chapters XII and XIII has been obtained to a very considerable extent from Admiral Shadwell's pages.

Shortland, Capt. R.N., Papers on Nautical Surveying, published in Naval Science, Vols. II., III., IV.
Simms. Treatise on Mathematical Instruments.
Simms. Treatise on Levelling.
Stevenson's Principles and Practice of Canal and River Engineering.
Villarceau and De Magnac. Nouvelle Navigation Astronomique. Paris, 1877.

Chapter I. of Part II. is devoted to the subject of Chronometers. M. Villarceau is a recognised authority in Chronometrical Science.

Note. A very important work on "Hydrographical Surveying" by Capt. Wharton, R.N., has been published while the sheets of the present Book were passing through the press. I should have been very glad to have read it while preparing my own materials. Its practical details and methods of proceeding by an officer of acknowledged capacity and wide experience are of the greatest interest, and its collection of Tables at the end must be of the highest value to all Practical Surveyors.

EXTRACT FROM THE REGULATIONS RELATIVE TO EXAMINATION FOR THE RANK OF LIEUTENANT.

NAUTICAL SURVEYING.

Use of Charts; Rating of Chronometers; Determination of Meridian Distance; Selection and Measurement of a Base Line; Determination of Latitude, Longitude, and True Bearing; Triangulation; Levelling; Soundings; Fixing Positions; Tide Gauge; Establishment of the Port. 100 marks.

INSTRUMENTS.

Construction and Use of Marine Barometer, Sextant, Artificial Horizon, Azimuth Compass, Theodolite and Level.
40 marks.

MARINE SURVEYING.

CHAPTER I.

SYMBOLS USED IN CHARTS AND SURVEYING.

I. Symbols used to denote the Quality of the Bottom. II. General Abbreviations. III. Buoys. IV. Lights and Lighthouses. V. Conventional Signs. VI. Soundings. VII. Miscellaneous. VIII. Surveying Symbols.

1. I. QUALITY OF THE BOTTOM:

Colour	b = blue.	blk = black.	br = brown.
	d = dark.	gn = green.	gy = grey.
	w = white.	y = yellow.	spk = speckled.
Substance	cl = clay.	crl = coral.	g = gravel.
	m = mud.	oys = oysters.	oz = ooze.
	peb = pebbles.	r = rock.	s = sand.
	sh = shells.	st = stones.	wd = weed.
Nature	brk = broken.	c = coarse.	f = fine.
	grd = ground.	h = hard.	rot = rotten.
	sft = soft.	stf = stiff.	

EXAMINATION.

Write down the abbreviations for the following Qualities of the Bottom:

Sand. Coarse gravel. Coral and fine sand. Pebbles mixed with shells. Broken shells. Stiff mud. Sand and pebbles. Grey mud.

What Qualities of the Bottom do the following abbreviations denote?

sh; st; r. s. sh; m; sft. cl; crl; stf. cl; c. s; gn. cl.; brk. sh.; wd; spk. sh; s; oz; gy. g.

2. II. GENERAL ABBREVIATIONS.

Alt = altitude. Anch^{ge} = anchorage. B = bay.
Bar = barometer. Bat^y = battery. B^k = bank.
C = cape. C. G. = Coastguard. Cath = Cathedral.
Ch = church. Chan = channel. Col^d = coloured.
Cr = creek.

E. D. = a *reported* Rock or Shoal whose *Existence* is *Doubtful*.

Fms = fathoms. Ft = feet. G = gulf.
G^t = great. H = hour. H^d = head.
Ho = house. H^r = harbour. H. W. = High Water.

H. W. F. & C. = High Water at Full and Change of the Moon.

I = island. I^s = islands. Kn = knot.
L = lake. Lat = latitude. Long = longitude.
L^t = light. L. W. = Low Water. Mag^z = magazine.
Mag^c = magnetic. M^t = mountain. Np = neaps.

Obs^n Spot + = observation spot. P = port.

P. D. = a Danger *known to exist* but its *Position* is *Doubtful*.

P^k = peak. P^t = point. R = river.
R^f = reef. R^k = rock. S^d = sound.
Sh = shoal. Sp = springs. Str = strait.
Tel = telegraph. Therm = thermometer. Var = variation.
Vil = village. W. Pl = watering place.

EXAMINATION.

Write down the abbreviations for the following terms:

Harbour, Hour, Head, House, River, Bank, Shoal, Magazine, Magnetic, Creek, Observation Spot.

What do the following abbreviations represent?

C. G.; Vil; L^t; M^t; P. D.; I; H^r; E. D.; G; Mag^c; S^d; Sp; Tel; Sh.

SYMBOLS USED IN CHARTS AND SURVEYING.

3. III. Buoys.

Colours. B. (near a Buoy) = Black.
Cheq. do. = Chequered.
H. S. do. = Horizontal Stripes.
R. do. = Red.
V. S. do. = Vertical Stripes.
W. do. = White.

Shapes.

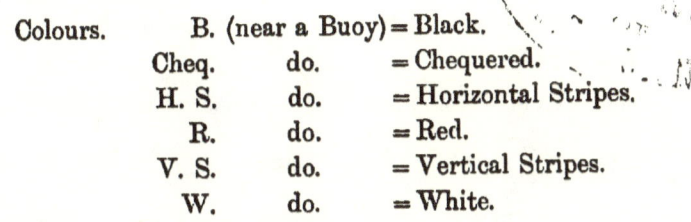

Can =   Nun =

Spar =   Conical, or Spiral =

Mooring =

Buoys with Beacons. The following Beacons are used on Buoys:—

Cage =   Triangle =

Diamond =   Globe =

The end of a Spit is usually marked thus:

4. Colouring Buoys. The following regulations are in use:—

CHEQUERED. The Buoy is divided into *four* horizontal and into *eight* vertical equal parts, to be coloured white and red, or white and black alternately. The white squares are then to be reduced by one inch all round.

VERTICALLY STRIPED. The Buoy is divided into *eight* equal parts, these being alternately white and red, or white and black. The white stripes are then to be made one-third narrower than the others.

HORIZONTALLY STRIPED. The Buoy is divided into *five* equal parts, these being coloured alternately white and black, or white and red. The white stripes are then made one-third narrower than the others.

These various colours are thus represented in charts:

White = (quite plain)   Red = (a little shading)

Black =   V.S. =   H.S. =

Cheq. =

5. REGULATIONS FOR THE EMPLOYMENT OF BUOYS.

(1) The side of a Channel is Starboard or Port with reference to a ship entering a Harbour *from seaward*.

(2) The entrance of a Channel, or a turning point, is marked by a conical or spiral buoy.

(3) The *Starboard* side of a Channel is marked with *Can* Buoys of the same colour throughout. Hence the well-known practical rule "whole-coloured buoys on the starboard side."

(4) The *Port* side of a Channel is marked with the *same shaped* buoys, but coloured vertically or chequered.

(5) If further distinction is required, then conical buoys with *globes* are on the *starboard*, and with *cages* on the *port*.

(6) When a shoal, or middle ground, exists in a channel, each end of it is marked by a buoy of that colour in use in that channel, but with *annular bands of white*.

If the shoal is of great extent, we must then proceed to use the buoys as though we had *two channels* to mark out, and the buoys must be placed in accordance with (3) and (4)

When necessary the *outer* buoy is distinguished by a *staff and diamond*, and the *inner* by a *staff and triangle*.

(7) The position of a wreck is marked by a *green* Nun Buoy.

EXAMINATION.

(1) Draw a *Can Buoy*, a *Nun Buoy*, a *Spar Buoy*, and a *Mooring Buoy*.

(2) Draw a *Chequered Nun Buoy*, a *Red Can Buoy*, a *Vertically striped Can Buoy*, a *Black Spiral Buoy*.

(3) Draw a Can Buoy with a Cage; a Nun Buoy with a Globe; and a Buoy used at the end of a Spit.

(4) You observe in a channel that all the buoys on your port side are black, are you entering or leaving the harbour?

(5) On going up a channel a buoy with annular bands of white is reported right ahead, what inference are you to draw?

(6) If a shoal in a channel is of great extent, how are the two extreme buoys distinguished from those placed at the sides?

(7) How is the position of a wreck marked?

(8) What rules are to be observed in painting a Chequered Buoy, a Vertical striped Buoy, and a Horizontal striped Buoy, respectively?

6. IV. LIGHTS AND LIGHTHOUSES.

The position of a Lighthouse is marked by a small round black dot, thus ⌀.

The following abbreviations are used:—

L^t F = Fixed Light.
F and Fl = Fixed and Flashing.
Lt Int = Intermittent Light.
Lt Occ = Occulting Light.
Min (near a Light) = Minutes.

L^t Fl = Flashing Light.
Fl^g L^t = Floating Light.
Lt Rev = Revolving Light.
Lt Alt = Alternating Light.
Sec (near a Light) = Seconds.

Vis (near a Light) = Visible.

7. EXPLANATIONS AND ILLUSTRATIONS.

1. F = FIXED OR STEADY. This term sufficiently explains itself. The great majority of the Lights on the British Coasts are of this description.

Examples: (1) Bishops' Rock in Scilly shews a White Fixed Light.

(2) At the end of the Breakwater in Brixham there is a Red Fixed Light.

(3) At the Pier Head of the Inner Harbour at Torquay there is a White Fixed Light to Seaward and a Red Fixed Light to the Westward.

(4) Southsea Castle shews a Fixed Light with Red and Green Sectors.

(5) Maplin Sands, a Red Fixed Light with a White Sector. The white indicates a particular channel.

2. FL = FLASHING. There are two kinds: (*a*) Flashes at short intervals; (*b*) Groups of Flashes at regular intervals.

In (*b*) the appearance presented consists of two or three flashes following each other as quickly as due separation allows, the remainder of the period separating these groups of flashes by a longer interval of eclipse.

Examples. (1) The Seven Stones Light Vessel shews three flashes in quick succession, followed by 36 seconds darkness. The whole period being one minute *.

(2) Casquets. Shews three successive flashes of about 2 seconds duration each, with intervals between each flash of about 3 seconds darkness, the third flash being followed by an eclipse of about 18 seconds.

(3) Royal Sovereign Shoal near Eastbourne. Shews three flashes in quick succession every minute; the time thus occupied is about 23 seconds, and then follows an eclipse of about 37 seconds.

(4) Galley Head (S. of Ireland). Shews six or seven flashes in quick succession every minute; the duration of this "group" of flashes is about 16 seconds, and then follows an eclipse of 44 seconds.

(5) Holyhead Breakwater. The Light at the end flashes every $7\frac{1}{2}$ seconds.

* The term "flash" must not be taken as denoting an instantaneous appearance, such for instance as that presented by Colomb's Flashing Signals; it is more deliberate than this, but quicker than the Revolving Light.

(6) The Arklow Light Vessel flashes twice in quick succession, followed by 45 seconds darkness. The period being a minute.

(7) Dungeness shews a white light which flashes every 5 seconds, the flashes being of about 2 seconds duration.

3. F. AND FL. = FIXED AND FLASHING. The appearance presented is a Fixed Light with the addition of White or Coloured Flashes preceded and followed by a short eclipse.

Example. Craigmore (Firth of Clyde). The period of the Light is about 11 seconds, and consists of 5 seconds light followed by three flashes of half a second each, these three flashes being separated by eclipses of $1\frac{1}{2}$, $\frac{1}{2}$, $1\frac{1}{2}$, $\frac{1}{2}$ seconds.

4. REV. = REVOLVING. The Light gradually increases to its full power, and then decreases to eclipse.

Note. At short distances, and in clear weather a faint continuous light may be seen.

Examples. (1) Wolf's Rock Light revolves every half minute.

(2) Start Light shews a White Light revolving every minute.

(3) Hanois Rock Light in Guernsey revolves every 45 seconds.

Note. It is deemed advisable that the light should not be obscured for too long an interval. There is now no Light in England withdrawn for more than 105 seconds, the whole *period* being 120 seconds. Thus the Light on Beachy Head revolves every two minutes, 15 seconds bright, and $1\frac{3}{4}$ minutes dark.

5. INT. = INTERMITTENT. A Light suddenly and totally eclipsed. The *brightness lasting for more than* 30 *seconds.*

Examples. (1) Tarbet Ness (E. of Scotland). White Light visible for $2\frac{1}{2}$ minutes, and eclipsed for $\frac{1}{2}$ minute.

(2) Ru Stoer. White Light visible for 1 minute, eclipsed for $\frac{1}{2}$ minute.

(3) River Ribble (W. of England). White Light visible for $3\frac{1}{2}$ minutes, eclipsed for $\frac{1}{2}$ minute.

(4) Dundrum Bay (E. of Ireland). Red Light visible for 45 seconds, eclipsed for 15 seconds.

(5) Rathlin (N. of Ireland). White Light with a Red Sector, bright for 50 seconds, eclipsed for 10 seconds.

6. Occ. = OCCULTING. A Light suddenly and totally eclipsed. The *brightness lasting less than* 30 *seconds.*

Examples. (1) West end of Plymouth Breakwater. The Light suddenly disappears for a space of 3 seconds every half minute.

(2) North Foreland Light suddenly disappears for the space of 5 seconds every half minute.

(3) Ardrossan Light, visible for 2 seconds, eclipsed for 2 seconds.

(4) Roche's Point (S. of Ireland), visible for 15 seconds, eclipsed for 5 seconds.

(5) Wicklow Light, bright for 10 seconds, dark for 3 seconds.

(6) Loop Head (W. of Ireland). A White Light visible for 20 seconds, and eclipsed for 4 seconds.

Note. The Intermittent and Occulting Lights are in contrast with Revolving Lights. In the first two the period of brightness is generally long and the period of eclipse is short, whereas in a Revolving Light the period of brightness is short and the period of darkness is long. Moreover in a Revolving Light there is a waxing and waning of the light as the beam packed, as it were, by the action of the annular rings of the lens approaches and recedes from the eye, and the place of the light is indicated (to a near observer) during the intervals of withdrawal by the reflections in the lantern; but in the case of both an Intermittent and Occulting Light the brightness is at its maximum throughout, and the intervening darkness is *total.*

7. ALT. = ALTERNATING. Red and White Lights alternately at equal intervals of time without any intervening eclipse.

Examples. (1) Owers Light Vessel shews alternately a *white* flash twice and then a *red* flash.

(2) Hartland Point (W. of England) shews alternately at intervals of half a minute a white flash twice and then a red flash.

8. ILLUMINATING APPARATUS.

Two systems are used:
The Catoptric* (C) or Reflecting system.
The Dioptric (D) or Refracting system.

In the Catoptric the light is usually produced by means of an Argand burner placed in the focus of a silvered metal reflector. In the Dioptric the light comes from one central lamp placed in the focus of a surrounding glass refractor, by which all the best rays of the light are sent horizontally, and to the surface of the sea.

A combination of the two is sometimes used, and is known as Catodioptric (C. D.), and where the light is very powerful it is possible to convey a portion of it (by reflection) to a lower aperture of the Lighthouse to guard against certain local dangers. Both systems admit of lights being shewn either as *fixed*, *intermittent*, *occulting*, *revolving*, *flashing*, or *group flashing*, and are classified as of various Orders; the Order being determined in the Catoptric system by the number of lamps and reflectors, and in the Dioptric by the size of the Instrument and of the central flame. Thus in a Dioptric 1st Order Light from the centre of the light to the vane is about 15 feet, in a 6th Order the vertical height is about 6 feet.

The Catoptric system is preferable where skilled labour is not obtainable, because less care and intelligence are required in the adjustment, it is also better in volcanic districts and in Light Vessels because less liable to be put out of adjustment.

The illuminant in each system is colza or paraffin oil or gas. The Electric Light is well adapted for the Dioptric system, the light being concentrated at a true focal point.

9. Coloured Sectors of light are produced by means of coloured glass placed inside the lantern.

It must be noted that the Bearings of Lights are always Magnetic and are given as *viewed from seaward*.

The Distances at which the Lights are visible are computed on the supposition that the observer's eye is 15 feet above the sea level.

* Katoptron, a mirror.

Light Vessels have usually the name of the danger they guard against painted distinctly on their sides. These vessels have as many Masts and Balls as they have Lights exhibited at night. When a Light Vessel has drifted from her correct position to one where she is useless as a guide the following method is adopted to intimate the fact. "A *fixed red light* will be exhibited at each end of the vessel, and a *red flare* shewn every quarter of an hour." In the daytime the Balls at the mastheads are struck.

When a Light Vessel marks the position of a wreck, the topsides are painted *green,* and in the daytime three Balls are placed on a yard 20 feet above the water, two of these vertically on the safe side, and one Ball on the danger side; in the night *fixed white lights* similarly arranged have the same signification.

When assistance is required from shore, special rockets of little sound but of great brilliancy are fired immediately after a gun.

When guns are used as fog signals they are fired at intervals of 10 minutes.

Sirens and horns in Light Vessels are always blown *against the wind.*

The Code of Signals in use is the International.

EXAMINATION.

(1) What two systems are used in lighting coasts? Specify the principle of each kind.

(2) Write down carefully the appearances presented to a mariner by the following descriptions of Lights: Lt F. L^t Fl. Lt F. and Fl. Lt Int. Lt Occ. Lt Rev. Lt Alt.

(3) I find near a Lighthouse marked on a chart the following note: "Rev. 20 sec vis 14 m." Explain it.

(4) What is a "Group Flashing" Light?

(5) What is the chief distinction between an Intermittent and Occulting Light?

(6) Contrast these with a Revolving Light.

(7) How does a Revolving Light differ from an Alternating Light?

I.] SYMBOLS USED IN CHARTS AND SURVEYING. 11

(8) How are Coloured Sectors produced? What is their chief use?

(9) In which system is the Electric Light used?

(10) Under what circumstances is the Catoptric System preferable to the Dioptric?

(11) On approaching land I see that a light is bright for 20 seconds and then suddenly disappears for 5 seconds. In what category is this Light to be placed?

(12) How are you to know by day and also by night whether a Light Vessel has drifted from her correct position?

(13) If a Light Vessel is placed near a wreck, how is she distinguished?

(14) How are you to know on which side the danger lies?

(15) In what way does a Lighthouse signal for help?

(16) In using a Foghorn what special precaution must be taken?

10. V. CONVENTIONAL SIGNS.

Sandy beach dry at L. W. =   *Note. Small dots* form the outside line, and behind these parallel rows of *finer dots* represent the sand.

Gravelly Beach =   *Small strokes* form the outside line, and behind these parallel rows of *dots* represent gravel.

Shingle or stony beach =   *Small zeros* form the outside line.

For examples of the above see the Plate in the Manual of Scientific Enquiry illustrating the Article on Hydrography.

Sandy shore =   *A sharp line* to signify the edge of the water at High Water.

Sand with gravel mixed (dry at L. W.) =   Closed curves of small dots to signify the sand with a few irregularly placed zeros and heavier dots to signify the gravel or coarse stones.

Sand Bank that dries at L. W. The same as the last but with the gravel or stones omitted.

Stone Bank and Beach =    Heavier dots in groups or single behind a continuous line.

Mud Beach (dry at L. W.) =   The faintest possible shading.

If *Banks*, not *Beaches*, are to be represented, then the continuous line at the edge is to be omitted.

Sand and Mud mixed (dry at L. W.) =    Very light shading with rows of dots to represent sand.

Coral Reefs =    Jagged outline with small crosses to denote rocks 6 feet below the water.

Rocky Ledges which cover and uncover =    Narrow rows of irregular shading, the darker parts signifying the sharper and higher ledges.

Rocks with less than 6 feet of water on them = (+)   Simply a cross within a limiting danger line.

Rocks awash at L. W. = (※)   A small cross with a dot in each quadrant. These dots may be considered as representing the *points of rock* appearing above the surface at L. W.

Isolated rocks which do not cover, i.e. which are visible at High Water Springs.

Rocks of different kinds with limiting danger line.

I.]   SYMBOLS USED IN CHARTS AND SURVEYING.   13

Cliffy Coast line = 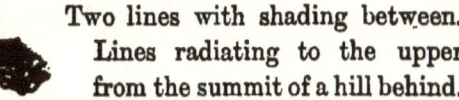   Two lines with shading between. Lines radiating to the upper from the summit of a hill behind.

Shore, steep to =   Two lines with dark shading between, the wider intervals signify that the height of the cliff increases.

Trees.   Mangroves

Palms

Pines

Generally

Swampy
Marshy } land =
Mossy

Grass
Meadow } land =

Cultivated grounds and gardens.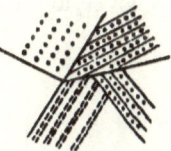

Kelp.

Note. "To economise the labour of the draughtsman, Towns and Buildings may be tinted *Red;* Sandy Beaches *Yellow;* Rocks shewing at Low Water *Brown;* Mud Banks *light Black;* Cultivated land *Green*.*"

EXAMINATION.

(1) Represent a Cliffy Coast line, and describe the method of doing so.

(2) How does a shore steep to differ from a cliffy coast line?

(3) What is the distinguishing difference in the modes of representing a Bank and a Beach?

(4) Draw the symbol for a Rock with less than 6 feet of water on it.

(5) Draw the symbol for a Rock awash at L. W.

(6) Draw a Ledge which covers and uncovers.

(7) Draw a Rock which does not cover.

(8) When is a Rock said not to cover?

(9) Draw a Coral Reef.

(10) How would you remember the difference between the symbols for a rock awash at L. W. and a rock with less than 6 feet of water on it?

(11) Draw a series of rocks with a limiting danger line.

(12) The limiting danger line is sometimes represented as a continuous line and sometimes by a dotted line, is any distinction sought to be thus conveyed?

11. VI. SOUNDINGS.

Shoal Banks which never uncover, and on which the depth of water is known

1 Fathom line	....	⎫	Sometimes when the depth
2 ,, ,,	.. ..	⎬	lies between 1 fathom and
3 ,, ,,	... ...	⎬	3 fathoms the depth is thus
4 ,, ,,	....:	⎭	marked

5 ,, ,,	....		Five dots in each series.
6 ,, ,,	....		Six dots in each series.

&c.

* General Instructions for Hydrographic Surveyors, p. 7, note.

I.]   SYMBOLS USED IN CHARTS AND SURVEYING.   15

10 fathom line —·—·—·—·— Strokes and single dots alternately.

20   ,,   ,,   —··—··—··— Strokes and two dots alternately.

30   ,,   ,,   —···—···—···— Strokes and three dots alternately.

100   ,,   ,,   ············   Continuous dots*.

$\overline{1\dot{3}\dot{0}}$, $\overline{\dot{5}\dot{0}\dot{0}}$ = No bottom found at the depth expressed.

12. VII. MISCELLANEOUS.

Breakers, Overfalls and Tide Rips 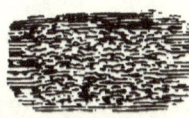 Long and short waving lines, such as would be used in representing a river rapid.

Thus the tidal stream in the Irish Sea encounters an extensive projection of the Codling Bank near Wicklow. The outer portion of the stream takes the circuit of the Bank, the inner stream sweeps over it, occasioning an overfall and strong rippling all round the edge by which the Bank may generally be discovered.

Tide Tables, p. 125.

Anchorage for large vessels ⚓ a complete anchor.

Anchorage for small vessels one fluke, or kedge.

Churches ✠

Stone windmill ✗ The building is round.

Wooden windmill ⚒ The building is high.

Villages A few houses scattered by the sides of the roads, and a few irregularly placed in the rear.

---

* In the Ordnance Maps of Great Britain the boundaries of Counties are represented by the 10 fathom line used in Charts, and those of Townlands by the 100 fathom line.

Towns The streets are represented by straight lines.

Currents ⟶ An arrow with feathers on both sides.
Flood tide stream ⟶ Feathers on only one side.
Note. The feathers are placed on that side on which the stream is inclined to press, if the arrow is thrown out of its line.
Ebb tide stream ⟶ Plain dart without feathers.

A *current* may be considered as the movement of a large body of water. A *stream* as the movement of surface water to the depth of about 70 fathoms.

High Water on Full and Change Days, i.e. on the days of Full and New Moon, is thus represented, H. W. F. & C. The Hour is always expressed in Roman figures, e.g. VII^h. 22^m.

The Compass on a chart is always drawn on the *magnetic* meridian, and hence if it is ever required to construct such a compass and to subdivide one quadrant, this quadrant must be taken between two cardinal points of the Magnetic Compass and not of the True.

The *underlined figures* on a bank signify the depth of water over it at H. W. or else the height of the bank above L. W.

The method adopted is of course fully explained in the Title of each Chart.

All Heights are given in feet above H. W. Ordinary Springs, and where there is no tide, then above the sea level.

All Bearings, as well as the directions of winds and currents, are *magnetic*.

In the Admiralty Charts, if changes have been made in the original survey of sufficient importance to require a *new plate* being engraved, the date of any such change is given at the bottom of the chart. Such a change necessitates the former chart being "destroyed in the presence of the Captain," but minor alterations (e.g. in buoys, lights, &c.), made in red ink by the Navigating Officers, are intimated by the dates in the lower left-hand corner of the chart. The official number of a chart is always found in the *right-hand lower corner*.

SYMBOLS USED IN CHARTS AND SURVEYING.

The periods of a tide are four in number, viz. 1st, 2nd, 3rd, 4th Quarters, and are denoted by 1st Qr., 2nd Qr., &c.

The *Velocity* of the tide is given in knots and fractions of a knot.

Thus the symbol ⟶ signifies a Flood tide stream
2nd Qr. 2½kn.
at the 2nd Qr. running due East at a rate of 2½ knots an hour.

A *Landmark* is some particular feature of a coast by means of which the locality may be known by a ship approaching or making the land, or by which she can find her way to an anchorage.

A *Day Mark* is a conspicuous object to mark a narrow entrance to a harbour or river, which from the configuration of the coast is difficult to make out from seaward.

A *Leading Mark* is generally denoted by *two very fine parallel lines* close together, and is required in order to keep in a certain channel or passage.

A *Clearing Mark* on the contrary is used to enable a vessel to keep clear of a certain danger, and is generally denoted on a chart by a *single dotted line*. Thus if a ship is beating up a passage wherein a danger exists, she will keep on a certain course until she gets two specified objects in one, or "in transit" as it is called, when she must alter course to avoid the danger ahead.

Thus we may have this note on a chart or in Sailing Directions: "Lighthouse in one with East Peak clears reefs in 5 fathoms, and kept open (S.ᵇE.) leads through the passage in Mid Channel."

EXAMINATION.

(1) Represent a 5-fathom line of soundings.

(2) Represent the 10-fathom, 20-fathom, and 100-fathom lines.

(3) "No bottom at 450 fathoms." Draw the representative symbol.

(4) Draw the symbols used to represent
  An anchorage for large vessels,
  An anchorage for small vessels,
  A church,
  A stone windmill,
  A wooden windmill.

(5) Draw the symbols which represent
  A current,
  A flood tide stream,
  An ebb tide stream,
  A flood tide stream setting E.S.E. at the 3rd quarter 5 knots.
  An ebb tide stream setting N.bW. at the 4th quarter $2\frac{3}{4}$ knots.

(6) What do the *underlined* figures on a bank signify ?

(7) A current sets N.E.; is this the True or Magnetic Bearing ?

(8) Distinguish between a Leading Mark and a Clearing Mark.

(9) How is each represented on a chart ?

(10) What is a Landmark ?

13. VIII. SURVEYING SIGNS.

⊙ = Observatory Station.

△ = Station where angles are taken either with the sextant or theodolite.

⟶ = R. T. = Right extreme, or Right Tangent.

⟵ = L. T. = Left extreme, or Left Tangent.

⊼ = B. M. = Bench Mark. This is the well-known mark used in the Ordnance Survey to point out the places where the levelling staves were placed in running the lines of levels. We shall have more on this subject below.

⌽ or ⌽ = Two objects in one, or in "transit."

⊖ = Altitude of the sun's centre.

⌽ = Bearing of the sun's centre.

⊙| = Right limb of the sun.

|⊙ = Left limb of the sun.

⊙ = Sun's lower limb.

⊙ = Sun's upper limb.

⊙ = Sun's lower and right limbs, &c.

Authorities for this Chapter:
    The Admiralty List of Abbreviations.
    General Instructions for Hydrographic Surveyors, 1877. This contains much information in the Appendices on the subject of Buoyage, Lighthouses, &c.
    "Tide Tables for the British and Irish Ports for the years 1881 and 1882."
    "Admiralty List of Lights in the British Islands," 1882.

Raper in his *Navigation* has a very good section on the employment of symbols. Vide pages 381—389. 3rd Edition.

CHAPTER II.

THE CONSTRUCTION AND USE OF SCALES.

14. Def. A Scale is an artificial means of representing any given dimensions, whether angular or linear. E.g. we may suppose that a mile is represented by a straight line 1 inch long, then 2 miles will be represented by 2 inches, and $3\frac{1}{2}$ miles by $3\frac{1}{2}$ inches, &c.

The different kinds of scales may be divided into two classes, viz. (I.) *Simply-divided Scales*, or *Plain Scales;* (II.) *Protracting Scales*.

15. (I.) Simply-divided Scales.

Of these there are three kinds, (α) *Scales of Equal Parts*, (β) *Diagonal Scales*, (γ) *Vernier Scales*.

(α). *Scales of Equal Parts* may be described as follows:—

Assume any convenient length, suppose 6 inches. Draw a straight line 6 inches long, and divide it into 6 equal parts. These six parts are called the *Primary* divisions. Next divide the first or left-hand *primary* into 10 equal parts, these smaller parts are known as the *secondary* divisions.

Hence we can see that the primary divisions will represent units, if the secondary divisions represent tenths; the primaries will represent hundreds if the secondaries are taken as tens, &c.

These scales of equal parts are generally contained between two fine parallel lines. The two lines of the first primary are divided, one into 10 equal parts, the other into 12 equal parts; one

CHAP. II.] THE CONSTRUCTION AND USE OF SCALES. 21

of the lines so divided will measure units and decimals, the second will measure feet and inches.

DEF. If the first primary be an inch long and be divided into 30 equal parts*, the scale is called a scale of 30. Similarly we can have scales of 40, of 50, &c. On a Protractor we have usually plain scales of 30, 35, 40, 45, 50, and 60 on one side, the number which denotes the scale standing on the left of the scale. On the opposite side of the Protractor we have scales of 1, $\frac{7}{8}$, $\frac{3}{4}$, $\frac{5}{8}$, $\frac{1}{2}$, $\frac{3}{8}$, $\frac{1}{4}$, $\frac{1}{8}$ inch, both decimally and duodecimally divided as explained above.

16. USE OF THESE SCALES. If a secondary division be taken as a unit, the primary will be 10; if the secondary be taken as 10, the primary will be 100; if a secondary be taken as $\frac{1}{10}$, the primary will be a unit, &c. Hence we can take off from these scales feet and decimals of a foot, yards and decimals of a yard, &c. Thus if the scale of a plan is $\frac{1}{2}$ inch to a mile, and we wish to lay off a distance of 3·6 miles, we must look for the scale on the Protractor with $\frac{1}{2}$ before it, and then placing one point of the compass on the line with 3 near it we must extend the other point to the 6th secondary division on the decimal line, and this interval will represent 3·6 miles.

17. MODE OF CONSTRUCTING THESE SCALES.

Let it be required to draw a plain scale, 10 units to the inch, and to exhibit 60 units.

Draw a line 6 inches long, and divide it into 6 equal parts. These are the *primary* divisions. Divide the first primary into 10 equal parts. Next draw a *thicker* line at a short distance ($\frac{1}{10}$ or $\frac{1}{18}$ inch) below the first line and draw vertical lines between them from the divisions on the first line as in the diagram.

Observe the position of the numbers. The zero is placed at the mark between the first and second primary divisions, and then from left to right in order are the numbers 10, 20, &c. This method of numbering is observed in all plain scales. The student

---

* Or what amounts to the same, if one-third of the inch be divided into ten equal parts, as on most protractors.

ought to examine the scales on his Protractor to impress this on his mind. In this way lengths are more easily taken off from scales; e.g. to take 15 units from the above scale, we must extend the points of the compass from 10 back to the 5th division in the subdivided primary.

When very minute divisions are not required we make use of these plain scales, because it is easy to subdivide an inch into 10, 15, or 20 equal parts, but if *hundredths* of an inch are required recourse must be had to a Diagonal Scale, as we shall see later on.

Exercises in Plain Scales.

(1) Construct a scale of 12 feet to 1 inch, and exhibit a length of 70 feet.

By proportion $12 : 1 :: 70 : x$; $\therefore x = 5·83$ inches.

Now draw a line 5·83 inches long*. Divide it into 7 equal parts, each of which will represent 10 feet. Subdivide the first primary into 10 equal parts, and each of these subdivisions will represent a single foot. Number as directed above, and print at the beginning "Scale of," and at the end of the scale print the word "feet"; we shall thus have a "Scale of......70 feet."

Note. Since 1 inch represents 12 feet, or 144 inches of real length, we have the "Natural Scale" $= \frac{1}{144}$.

Def. The Natural Scale is the ratio that the length of a certain unit on the paper or plan bears to the real length of that unit on the earth's surface.

According to the method described above all the Plain Scales are constructed; the *secondary* divisions varying only with the number of units represented by a primary division; e.g. if we wish to shew feet and inches, a primary will represent 1 foot, and the first primary must be divided into 12 equal parts. If fathoms and feet are to be represented, then each primary will represent a fathom, and the first primary must be divided into 6 equal parts. If furlongs and chains, the first primary must be divided

---

* This length must be taken from a Diagonal Scale.

II.]   THE CONSTRUCTION AND USE OF SCALES.   23

into 10 parts (because a chain = 22 yards, and a furlong = 220 yards). If miles and furlongs, the first primary will contain 8 equal parts. If miles and cables, the first primary must contain 10 equal parts.

(2) Construct a scale of 20 yards to 1 inch, to shew yards and feet, and exhibit a length of 90 yards.

(3) Construct a scale of 10 miles to 1 inch, to shew furlongs, and exhibit 50 miles.

(4) Subdivide into cables a scale on which 11·80 inches are equal to a sea mile, and shew what length will represent 1000 feet.   (March, 1875.)

(5) Subdivide into cables a scale on which 7·3 inches are equal to a nautical mile, and shew on the scale a space of 500 feet.   (April, 1875.)

(6) Subdivide into cables a scale on which 10 inches are equal to half a nautical mile, and shew on the scale the lengths of 100 feet and 1000 feet.   (Dec. 1875.)

(7) Define the term Natural Scale as used on plans.
   (June, 1876.)

(8) Subdivide into cables a scale on which 6 inches are equal to a mile of latitude: draw also the corresponding scale of a mile of longitude, the latitude being 50° N.   (June, 1876.)

(9) Draw the corresponding scale in the last question if the latitude is 45° N.   (Dec. 1876.)

(10) Subdivide into cables a scale on which 6·3 inches are equal to a mile of latitude in latitude 35° N., and draw also the corresponding scale of a mile of longitude.   (May, 1877.)

(11) Draw a straight line 4·75 inches long, and divide it into 10 equal parts.   (Oct. 1877.)

(12) How would you test the accuracy of scale of the published plan of a harbour in which your vessel was lying?
   (Beaufort, March, 1880.)

(13) The scale of a plan being 5 inches to a nautical mile of 6080 feet, calculate the length on the scale for the 10 feet pole

(used in putting in coast line) corresponding to an observed angle of 6′ subtended between the cross-bars*. (June, 1880.)

(14) On a plan the scale is 2 inches to a mile. Calculate the natural scale†. *Result* $\frac{1}{31680}$.

(15) Find the natural scale when a mile is represented by an inch. *Result* $\frac{1}{63360}$.

(16) Find the natural scale when 4 inches represent a mile. *Result* $\frac{1}{15840}$.

(17) Find the natural scale when 1 foot represents a mile. *Result* $\frac{1}{5280}$.

(18) Given the natural scale $\frac{1}{10560}$, find the scale on the chart.

Let $x =$ number of inches which represent a mile,

$$\therefore \frac{x}{5280 \times 12} = \frac{1}{10560}, \quad \therefore x = \frac{5280 \times 12}{10560} = 6,$$

∴ 6 inches represent a mile.

(19) Given the natural scale $\frac{1}{1013760}$, find the length which 1 inch represents. *Result* 1 inch = 16 miles.

(20) Given the natural scale $\frac{1}{253440}$, what length represents a mile? *Result* ¼ inch.

(21) Given the natural scale $\frac{1}{3960}$, what length represents a mile? *Result* 16 inches.

(22) Given the natural scale $\frac{1}{2500}$, what length represents a mile? *Result* 25·34 inches.

18. *Note.* In the Ordnance Survey of Great Britain the following scales are adopted:

(1) Towns = $\frac{1}{500}$, or 126·72 inches to the mile.
(2) Parishes = $\frac{1}{2500}$, or 25·34 „ „ „
In this scale, a square inch represents an acre.
(3) Counties = $\frac{1}{10560}$, or 6 inches to the mile.
(4) The Whole Kingdom = $\frac{1}{63360}$, or 1 inch to the mile.

* See Chapter VI. for a description of this Instrument and the method of computing the Scale.
† The mile in this and the following questions = 5280 feet.

A natural scale of $\frac{1}{50000}$ will permit every hill 100 feet high, a pond 100 feet in diameter, woods 200 or 300 feet across, towns, and large isolated buildings to be delineated. A natural scale of $\frac{1}{10000}$ will permit of a complete topographical representation of every object in a country (except fences) in exact proportion to the extent it occupies.

19. (β). DIAGONAL SCALES.

In the plain scales as already explained we can subdivide an inch into 10 or even 20 equal parts without difficulty, but when it is necessary to take off a length to two decimal places, i.e. to hundredths, we must make use of a diagonal scale. These scales are constructed as follows:—

Draw *eleven* straight lines parallel to one another and $\frac{1}{10}$ of an inch apart. Divide the uppermost of these lines into equal parts, these primary divisions being of any required length such as an inch, a half-inch, &c. Through these points draw perpendiculars cutting all the parallels, and number these primary divisions beginning at the extreme left thus 1, 0, 1, 2, 3,... as already explained in the construction of the plain scale.

Next subdivide the top and bottom lines of the first primary into 10 equal parts, and number these alternate divisions 2, 4, 6, 8 *from right to left* along the bottom line; and number the alternate parallels 2, 4, 6, 8 *from the bottom upwards*. Then draw lines as in the diagram, viz. from the zero of the bottom line to the first division of the top, from first division of bottom to the second of the top, &c., and the scale is completed.

PROOF. In the triangles AX_1Y_1 and ABC_1, by Euclid VI. 4, we have

$$\frac{X_1Y_1}{BC_1} = \frac{AX_1}{AB} = \frac{1}{10}, \quad \therefore X_1Y_1 = \frac{1}{10}BC_1; \text{ but } BC_1 = \frac{1}{10} \text{ inch,}$$

$$\therefore X_1Y_1 = \frac{1}{10} \cdot \frac{1}{10} \text{ inch} = \frac{1}{100} \text{ inch.}$$

Now if a compass be extended along the second parallel from the bottom, from the perpendicular marked 2 to the point Y_1, it will take off a distance equal to 2 inch + ·01 inch = 2·01 inch.

Again $\dfrac{X_2Y_2}{BC_1} = \dfrac{AX_2}{AB} = \dfrac{2}{10}.$ $\therefore X_2Y_2 = \dfrac{2}{10}BC_1 = \dfrac{2}{100}$ inch.

∴ from perpendicular marked 2 to Y_2 = 2·02 inches.

Hence if we wish to take off 1·1 inch we measure ZC_1,

.... 1·2 .... ZC_2,

.... 1·3 .... ZC_3,

&c.

Similarly, we can see that the interval from the perpendicular marked Z to the point marked P will be 1·75 inches.

Note. To take from a diagonal scale the number

5·74, we must consider the primary divisions as *units*,

46·70, ... *tens*,

253·00, ... *hundreds*.

EXERCISES IN THE CONSTRUCTION OF DIAGONAL SCALES.

(1) Construct a scale of 120 feet to an inch to measure 700 feet, and from which single feet may be taken.

It is evident that we cannot construct a plain scale to answer the purpose, because we could not divide an inch into 120 equal parts. The scale required must therefore be a diagonal scale, and this is constructed as follows:

As 120 : 1 :: 700 : x. ∴ x = 5·83 inches.

Draw a straight line 5·83 inches long; divide this into 7 equal parts; each part will represent 100 feet. Subdivide the first primary into 10 equal parts, therefore each secondary will represent 10 feet. Now draw 10 lines parallel to the first line and $\frac{1}{10}$ inch apart; draw the diagonal lines as directed above. Then we can take off distances to single feet.

(2) Construct a scale of 9 miles to 1·3 inches to measure 40 miles, and from which distances to furlongs may be taken.

$$\text{As } 9 : 1\text{·}3 :: 40 : x. \quad \therefore \ x = 5\text{·}77 \text{ inches.}$$

Divide a line 5·77 inches long into four parts, each part will represent 10 miles. Divide the first primary into 10 parts, each will represent a mile. Next draw 8 lines parallel to the first, or scale line: draw the diagonal lines, and it is evident that we can read to furlongs.

(3) Construct a scale of 10 feet to 1·5 inches, to measure 40 feet, and from which distances to inches may be taken.

$$\text{As } 10 : 1\text{·}5 :: 40 : x. \quad \therefore \ x = 6 \text{ inches.}$$

Divide a line 6 inches long into four parts, each will represent 10 feet. Subdivide the first primary into 10 parts, each will represent a foot. Next draw 12 lines parallel to the scale line. Draw the diagonal lines as usual, and it is evident that we can take off distances to inches.

(4) Construct a scale of 100 fathoms, with 18 fathoms represented by 1 inch, and from which we can measure feet.

$$\text{As } 18 : 1 :: 100 : x. \quad \therefore \ x = 5\text{·}55 \text{ inches.}$$

Divide a line 5·55 into 10 parts, each will represent 10 fathoms. Subdivide the first primary into 10 parts, each will represent 1 fathom. Now draw 6 lines parallel to the scale line, and then we can take off single feet.

(5) Construct a scale of 76 miles to 1·3 inches to read to single miles, and to exhibit 500 miles.

$$\text{As } 76 : 1\text{·}3 :: 500 : x. \quad \therefore \ x = 8\text{·}55.$$

Divide 8·55 inches into 5 parts, each will be equal to 100 miles. Divide the first primary into 10 parts, each will represent 10 miles. Draw 10 lines parallel.

Exercises in the Construction of Scales.

(6) On a map 18 miles are represented by 3·7 inches. Complete the scale to 30 miles.

As $18 : 3·7 :: 30 : x$. ∴ $x = 6·17$ inches.

Divide 6·17 inches into 3 parts, each of which will be equal to 10 miles. Subdivide the first primary into 10 parts, each of these will represent a mile.

(7) On a map 30 miles are represented by 18·6 inches, draw a scale to exhibit 10 miles, and which will measure furlongs.

As $30 : 18·6 :: 10 : x$. ∴ $x = 6·2$ inches.

Divide 6·2 inches into 10 parts, each will represent a mile; subdivide the first primary into 8 parts, each will represent a furlong.

(8) A length of 4 ft. 6 in. is represented by a length of 2·5 inches. Complete the scale to 10 feet, such that we may measure inches.

$4·5 : 2·5 :: 10 : x$. ∴ $x = 5·55$ inches.

Divide 5·55 inches into 10 parts, each will represent a foot. Subdivide the first primary into 12 parts, each will represent an inch.

(9) If 3 inches represent 9 fathoms, complete the scale to 20 fathoms, such that single feet may be measured.

(10) Construct a scale of yards to shew 30 yards, the natural scale being $\frac{1}{187}$.

Here 1 inch = 187 inches = 15·58 feet = 5·19 yards.

As $5·2 : 1 :: 30 : x$. ∴ $x = 5·78$ nearly.

Divide a line 5·78 inches long into 3 parts, each will represent 10 yards. Subdivide the first primary into 10 parts, each of which will represent a yard.

(11) Construct a scale of 60 miles to shew miles, the natural scale being $\frac{1}{614373}$.

Here 1 inch = 9·7 miles. Length of scale = 6·18 inches.

THE CONSTRUCTION AND USE OF SCALES.

(12) Construct a scale of 100 feet to shew feet, the natural scale being $\frac{3}{500}$.

Here 3 feet = 500. Length of scale = 7·2 inches.

(13) Draw a scale shewing feet and inches, where the natural scale is $\frac{1}{12}$.

(14) Draw a scale shewing yards and feet, the natural scale being $\frac{1}{36}$.

(15) Draw a scale shewing miles and furlongs, the natural scale being $\frac{1}{63300}$.

(16) Draw a scale of 160 fathoms to 9 inches to shew feet.

(17) Draw a scale of 60 yards to 4·57 inches to shew feet.

(18) Draw a scale of 87 fathoms to 5·1 inches to shew fathoms.

(19) Draw a diagonal scale of 1 foot to ·87 inch, shewing inches.

(20) Draw a diagonal scale of 2 fathoms to ·55 inch, shewing feet.

(21) On a map 183 miles occupy a length of 14·3 inches, construct a scale to shew furlongs.

20. If it is desired to take from an ordinary 6-inch Protractor a distance of 9·92 inches, we must take the distance 9·92 from the ½-inch diagonal scale, and then "step" it twice. If we required 16·76 inches, take the distance from the ¼-inch scale, and "step" it four times.

21. (γ). VERNIERS.

The important scales which we are about to describe are named from their inventor, Peter Vernier, who died about 1637.

The Vernier is a contrivance for subdividing to any extent the smallest division on a graduated scale.

Verniers are either straight or curved. We have examples of the former in those attached to Barometers, and of the latter in those attached to the Sextant, Theodolite, Station Pointer, &c. The *principle* is the same whatever be the form.

22. Suppose a scale is divided into half degrees, i.e. into 30′ spaces, and let a length of 29 of these spaces be taken from the scale by means of a pair of compasses, and placed on paper. Then if this space thus transferred be divided into 30 equal parts, it is evident that *each* of these latter parts $= \dfrac{29 \times 30'}{30} = 29'$.

Now 29′ is less than 30′ by 1′, i.e. by $\frac{1}{30}$ of 30′.

23. Now let n divisions of the Vernier $= (n-1)$ divisions of the limb.

Let $l =$ value of one division of the limb,
and $v =$ value of one division of the Vernier;

$\therefore\ nv = (n-1)l\,;$

$\therefore\ v = \dfrac{n-1}{n} \cdot l\,;$

$\therefore\ l - v = l - \dfrac{n-1}{n} \cdot l = l\left(1 - \dfrac{n-1}{n}\right) = l\left(\dfrac{n-n+1}{n}\right) = \dfrac{1}{n}l.$

Hence the difference between a limb division and a Vernier division is $\dfrac{1}{n}$ th of the value of a limb division.

Note. This difference is known as the *Least Reading** of the Vernier, and expresses the degree of minuteness to which we are enabled to read by its aid.

24. We may discuss the method of construction in a general manner as follows:—

Let $l =$ value of a limb division,
and $n =$ number of divisions on the Vernier.

Let $(pn - 1)$ limb divisions be taken as the length of the Vernier, and let this be divided into n equal parts.

Then we have $\quad nv = (pn - 1)l\,;$

$\therefore\ v = \dfrac{pn - 1}{n} \cdot l\,;$

---

* Other terms to express this difference are "Accuracy of Reading," and "Least Count."

THE CONSTRUCTION AND USE OF SCALES.

$$\therefore pl - v = pl - \frac{pn-1}{n}.l$$

$$= \frac{pnl - pnl + l}{n}$$

$$= \frac{l}{n}.$$

Now it is evident that the value of p does not affect the value of the Least Reading (which under all circumstances is the nth part of l), but by taking p as 2 or 3 we gain the advantage of being able to distinguish without difficulty the marks in exact coincidence.

In the case of the sextant, n is generally 60 and l is 10′, hence we must have the Least Reading $\frac{10 \times 60}{60} = 10''$. From the above discussion it is evident that we can take as the length of the Vernier $(2 \times 60 - 1)$ limb divisions. This, therefore, explains why 60 Vernier divisions are equal to 119 limb divisions.

25. It is usual, but by no means necessary, to have the divisions of the Vernier *smaller* than those of the limb. Let the Vernier divisions be the *greater*. Then

n divisions of Vernier $= (n+1)$ divisions of limb;

$$\therefore nv = (n+1)l;$$

$$\therefore v = \frac{n+1}{n}.l;$$

$$\therefore v - l = \frac{n+1}{n}.l - l = l\left(\frac{n+1}{n} - 1\right) = l\left(\frac{n+1-n}{n}\right) = \frac{1}{n}l.$$

Hence the Least Reading is as before $\frac{1}{n}$th of a limb division.

The only practical difference which this will make is that in ordinary instruments where the graduations proceed from right to left, *we shall have to count from left to right*, i.e. the zero of the Vernier must be considered as being situated at the extreme left instead of at the extreme right.

EXERCISES.

(1) If a sextant limb be divided into 10′ spaces, and 119 of these be taken for the length of the Vernier, and the Vernier contain 120 parts, find the Least Reading.

Result $\frac{1}{120}$ of $10' = 5''$.

(2) If a limb be divided to 10′, and 59 of these be taken for the length of the Vernier, and this space on the Vernier be divided into 60 parts, to what minuteness may be read off?

Result $\frac{1}{60}$ of $10' = 10''$.

(3) Suppose that an inch is divided into 10 equal parts, and a Vernier be constructed equal in length ·9 inch, and this Vernier space be divided into 10 parts, find the Least Reading.

Result $\frac{1}{10}$ of $\frac{1}{10}$ inch $= \frac{1}{100}$ inch.

(4) If 29 spaces of 10′ be divided into 30 parts, find the Least Reading. *Result* $\frac{1}{30}$ of $10' = 20''$.

(5) If 29 spaces of 20′ be divided into 30 parts, find the Least Reading. *Result* $= 40''$.

(6) If 29 spaces of 30′ be divided into 30 parts, find the Least Reading. *Result* $= 1'$.

(7) If 39 spaces of 30′ be divided into 40 parts, find the Least Reading. *Result* $= 45''$.

(8) If 19 spaces of 1° be divided into 20 parts, find the Least Reading. *Result* $\frac{1}{20}$ of $1° = 3'$.

26. *Given the Least Reading, and the graduations of the limb, to construct the Vernier.*

Here we have evidently to find n, i.e. *the number of spaces on the Vernier*, and then the number of spaces on the limb which must be taken, *is one less or one more* than any multiple of n.

We know from what has gone before that

$$l - v = \frac{1}{n} l;$$

$$\therefore n = \frac{l}{l - v} = \frac{\text{value of a limb division}}{\text{Least Reading}}.$$

II.] THE CONSTRUCTION AND USE OF SCALES. 33

Of course l, and $l-v$ must be expressed in the *same unit*, minutes, inches, feet, &c.

Exercises.

(1) Construct a vernier which shall enable us to read to 10″, the limb being divided to 10′.

Here $$n = \frac{l}{l-v} = \frac{10 \times 60}{10} = 60.$$

Hence we must take 59 or 61 divisions of the limb as the length of the vernier, and then this space must be divided into 60 parts.

(2) Construct a vernier which shall enable us to read to ·01 inch, the inch containing 10 parts,

$$n = \frac{\frac{1}{10}}{\frac{1}{100}} = 10.$$

Hence we must take ·9 or 1·1 inch and divide it into 10 parts.

(3) Suppose a limb is divided into 20′ spaces, construct a vernier which shall enable us to read to minutes.

Result. 20 spaces on vernier = 19 or 21 on limb.

(4) Suppose we wish to read to 30″ in the last example.

Result. 40 on vernier = 39 or 41 on limb.

(5) Suppose we wish to read to 3″ on a reflecting circle which is graduated to 5′.

Result. 100 on vernier = 99 or 101 on limb.

(6) The arc of a sextant is divided to 10′. If 119 of these arc divisions be taken for the length of the vernier; into how many divisions must the vernier be cut to give readings to 5″ and 10″? *Result.* 120 and 60.

(7) If the arc of a sextant be divided to 10′, and 79 of these divisions are taken, and this space is divided on the vernier into 40 equal parts; to what extent may readings be obtained?

Result. 15″.

(8) Make a vernier to read 15″ on a sextant which is divided to 15′. *Result.* 60 vernier = 59 or 61 limb.

(9) Make a vernier to read to 20″ on a sextant divided to 15′.

Result. 45 vernier = 44 or 46 limb.

R. M. S.

(10) The difference between two arc spaces and one vernier space is 15″, and the vernier is divided into 60 equal parts. What is the value of a limb division ? *Result.* 15′.

(11) If m divisions on limb are equal to 1°, if n of these are taken to form a vernier, and the vernier is divided into p equal parts, find the Least Reading. (May, '80).
Result. $\dfrac{3600''}{mp}$.

(12) The limb is divided to 10′, and m divisions on the limb are equal to n divisions on the vernier; find the Least Reading (June, '80). *Result.* $\dfrac{600''}{n}$.

(13) If the degrees on the limb are divided into $5m$ equal parts, find the Least Reading, when the length of the vernier is equal to $2x$ of the limb divisions. *Result.* $\dfrac{360''}{mx}$.

(14) If a limb division is $\dfrac{mx}{y}$ degrees, and $pn-1$ of these are taken as the length of the vernier, and this length is divided into pn parts; find the Least Reading. *Result.* $\dfrac{3600''mx}{pny}$.

27. Principle of Reading Off.

Let the index of the vernier fall between two of the marks on the limb, between, e.g., x and $x+1$, reckoned from the zero of the limb, and let $l=$ value of a division on the limb.

Therefore whole reading $= xl +$ fraction from x to the index of the vernier.

To discover the value of this fraction we note that the mth division of the vernier coincides with a division on the limb; therefore the fraction $= m(l-v) = m \cdot \dfrac{1}{n} l = \dfrac{m}{n} l$.

Example. If the vernier be divided into 10 parts occupying a space of 9 divisions on the limb, and if the 4th division of the vernier coincides with a division of the limb, the whole reading $= xl + \tfrac{4}{10} l$.

II.]   THE CONSTRUCTION AND USE OF SCALES.   35

Suppose $l = 10'$, and x is the 74th division from the zero of the limb, then the reading $= 74 \times 10' + \frac{4}{10} \times 10'$
$$= 740' + 4' = 744' = 12° \ 24' \ 0''.$$

In practice we can always discover the value of xl from the numbers engraved on the limb, and the value of $\frac{m}{n} l$ from the numbers on the vernier.

Suppose then that in an ordinary sextant, the index of the vernier falls between $56° \ 10'$ and $56° \ 20'$, and that the 5th stroke to the left of 3 on the vernier coincides with a stroke on the Limb, we have evidently $m = 23$, therefore the fraction
$$\frac{m}{n} l = \frac{23}{60} \times 10' = 3' \ 50'';$$

$\therefore$ whole reading $= 56° \ 10' \ 0'' + 3' \ 50'' = 56° \ 13' \ 50''$.

Note. As already explained, the value of this fraction can be easily obtained by inspection; we find in this last case that the 5th stroke to the left of the 3rd long line on the vernier is the coincident division, and each smaller stroke measures $10''$, $\therefore$ the fraction $\qquad = 3' \ 50''$.

28. Generally, we may reason thus:—Let 29 divisions on the limb, each of half a degree, be equal in length to 30 on the vernier, then each vernier division is less than a limb division by $\frac{1}{30}$ of a limb division, i.e. by $1'$. Now, if the index of the vernier coincide with the zero of the limb, we shall have the 30th stroke on the vernier coincident with the 29th division on the limb. We shall have also the 1st division of the vernier $1'$ to the *right* of the 1st division of the limb*, the 2nd division of the vernier $2'$ to the right of the 2nd division of the limb, &c., and the 30th division of the vernier $30'$ to the right of the 30th division of the limb, and therefore coinciding with the 29th division of the limb, as it ought. Now, let the vernier be moved to the left until its 1st division (*not its Index*) coincides with the 1st division of the limb, then from what has been said, it is evident that the index of the vernier has moved $1'$; again, if the

* The graduation is supposed to proceed from right to left as in a sextant.

2nd vernier division coincide with the 2nd limb division, the index has moved 2', because originally this 2nd vernier division was 2' distant from the 2nd limb division, &c.

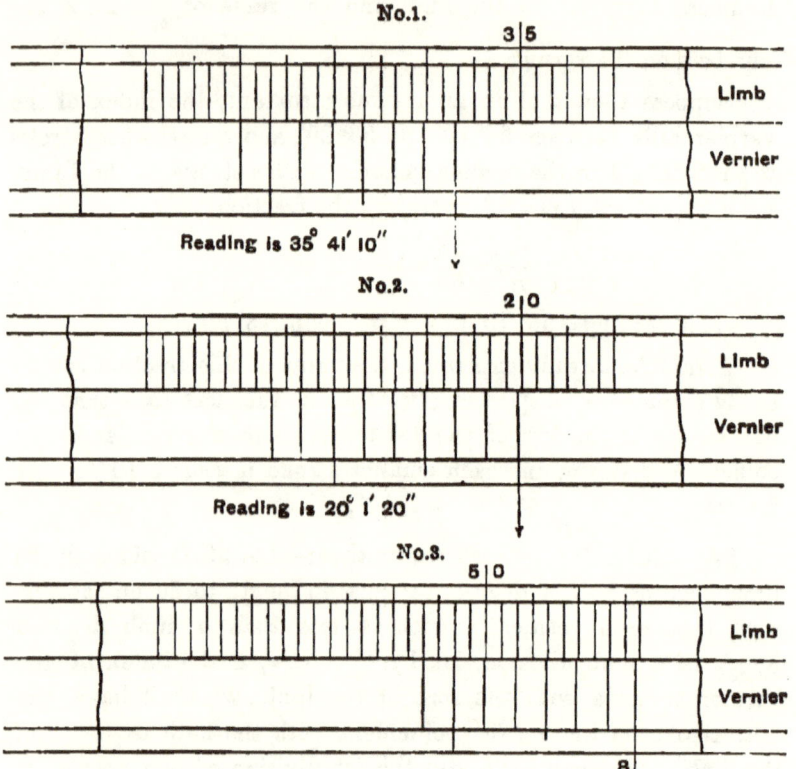

DIAGRAMS SHEWING READINGS ON A SEXTANT.

The Index is between 32° 20' and 32° 30'.
The Reading by the vernier is 9° 10', ∴ the whole angle observed is 32° 29' 10".

29. To DETERMINE WHETHER THE LIMB IS CORRECTLY DIVIDED, we make the index of the vernier coincide with some stroke on the limb, then if the last vernier stroke coincide with a stroke of the limb, and this occurs *at all parts* of the limb, we may assume that the graduations are accurate.

Suppose that the coincidence occurs at x divisions (reckoned on the vernier) from the last, then the vernier is too long or too short by x times the least reading; this should be ascertained in several places and the mean result taken: let the mean error $= e$, then $(n-1)l = nv + e, \therefore n(l-v) = l + e, \therefore l - v = \dfrac{l}{n} + \dfrac{e}{n}$.

Hence a reading in which the fraction is $m(l-v)$ now becomes

$$\dfrac{m}{n}l + \dfrac{m}{n}e.$$

The correction is $+ m\dfrac{e}{n}$ if the vernier is too short by e, and

$- m\dfrac{e}{n}$ if the vernier is too long by e,

e.g. If the limb is divided to 10', and the vernier gives 10" as the least reading (in which case $n = 60$), and the vernier is found too short by 5", then the correction is $m \cdot \dfrac{5''}{60}$: but every 6th division on the vernier gives 1', we must therefore add 0·5" for every minute read on the vernier.

30. II. Protracting Scales.

Of these there are several kinds. We shall first give the names of these scales, and the abbreviations by which they are usually denoted on Protractors and Sectors, then intimate the uses to which they are put, and finally enter into more detail as regards the construction and method of using the scales which are of most importance in connection with our subject.

The chief Protracting Scales are as follows:

  (1) Chords denoted by Cho. and C.
  (2) Rhumbs   ,, Rhu.
  (3) Latitudes   ,, Lat.
  (4) Longitudes ,, Lon.
  (5) Sines   ,, Sin.
  (6) Secants   ,, Sec.
  (7) Tangents   ,, Tan.
  (8) Semi-tangents ,, S. T.
  (9) Hours   ,, Hou.
  (10) Polygons   ,, Pol.

31. USES OF THESE SCALES.

The scales of "Latitudes" and "Hours" are used in the construction of sun-dials.

The scales of "Sines," "Secants," "Tangents," and "Semi-tangents" are used in the various projections of the sphere.

The scale or line of "Polygons" is useful in inscribing a regular polygon in a circle, or describing a regular polygon on a given straight line.

The scale of "Longitudes" shews the number of equatorial miles in a degree of longitude at the parallel of latitude indicated by any degree on the scale of chords which is usually placed next to it. E.g. In latitude 60°, a ship sails E. 79 miles, find the D. Long. made good. Opposite 60 on the scale of chords is 30 on the scale of longitudes. This is the number of equatorial miles in a degree of longitude at that latitude;

$$\therefore \text{ as } 30 : 79 :: 60 : x;$$

$\therefore x = 158$ miles, the required D. Long.

The scale of "Chords" enables us to lay off an angle of any exact number of degrees, and to measure to any exact degree an angle already protracted.

The scale of "Rhumbs" is merely a scale of chords of the points and quarter-points of one quadrant of the compass, and is used in laying down the ship's course expressed in points, and in measuring a course already projected.

32. CONSTRUCTION AND METHOD OF USING A SCALE OF CHORDS.

The object of this scale, as already explained, is to lay down an angle from a given point in a given straight line, and to measure an angle already laid down.

The principle of its construction depends on the fact that the side of a regular hexagon inscribed in a given circle is equal to the radius of the circle (Euclid, B. IV. 15).

Draw two lines OA, OB at right angles; describe the arc AB with centre O and radius OA.

With centres A and B respectively and radii $AO = BO$, describe circles cutting the arc AB in the points marked 6 and 3 respectively. Trisect the arcs $A3$, 36, $6B$, and mark the points 1, 2, 3, ... 8. Draw the chord AB, and then with centre A

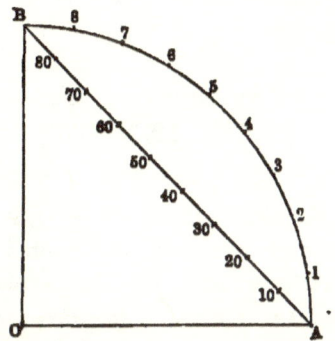

transfer to the chord AB, the distances $A1$, $A2$, $A3$, ... $A8$. Number these points 10, 20, ... 80. Then AB thus divided will be a scale of chords.

This scale is only divided to 10 degrees. If the quadrant be divided to degrees, and the distances from A be transferred as explained, we shall have a scale of chords divided to degrees, &c.

33. Use of the Scale of Chords.

Construct an angle of 45° by the scale of chords. Draw a straight line AB. From the scale of chords take the length from 0 to 60 and lay this distance off along AB; then describe an arc BD with A as centre, and this length AB as radius. Next take off from the same scale the length from 0 to 45, and describe the arc BC cutting the arc BD in the point C. Join AC, and the angle $BAC = 45°$.

To measure an angle by the scale of chords.

Let XAY be an angle, and it is required to find its value.

With A as centre and distance AB (0 to 60 from the scale) as radius describe an arc BC, cutting AX in B and AY in C. Take off the length of the chord BC from the paper, and placing one point of the compass at the zero of the scale of chords notice to what division on the scale the other point reaches, suppose 38, then the angle XAY is an angle of 38°.

It is easy to estimate by the eye to a quarter of a degree. If the angle is greater than 90 degrees; divide it by 2, and step off the distance twice; or, lay off 90 first, and then the excess of the angle over 90.

Examples. Lay off by means of the scale of chords the following angles:—40°; 73°; 107°; 136°; 163½°, and test the accuracy of your work by the graduated edge of your protractor.

The scale of "Rhumbs" is merely a scale of chords of the angles of deviation from the meridian denoted by the points and quarter-points of the compass. By its aid we can lay down a ship's course expressed in points, or measure a course already laid down.

E.g. Lay off a course N. N. E. ¾ E. by the scale of rhumbs.

Let A be the ship's position on any meridian. With AB (= 60 from the scale of chords attached) as radius describe an arc BD. Take off $BC = 2\frac{3}{4}$ from the scale of rhumbs. Measure off BC on the arc BD as already explained. Join AC, then $BAC = 2\frac{3}{4}$ points. By the converse process a course can be measured.

34. The Sector.

It seems advisable to describe this useful instrument in some detail, because in the hands of one who is well acquainted with its use, it is capable of giving many satisfactory results.

It is composed of two pieces of ivory, 6 inches long and ¾ inch wide, joined by a hinge at one end; this hinge allows the two limbs to be opened to any required extent.

On examining the instrument when opened fully, we find the following scales engraved on it. On the back, the length of a foot is divided into 10 primary parts, and these parts are again subdivided into 10, so that we have a foot divided into 100 *equal parts*. Close to one edge of the instrument we have 12 inches, and these inches decimally divided, so that we have the foot divided into 120 *equal parts*.

THE CONSTRUCTION AND USE OF SCALES.

On one face of the instrument we have 3 scales, marked respectively T, S, N. These are known as "Gunter's Lines," and are the scales of the logarithms of tangents, sines and numbers. It seems unnecessary to dwell more on these scales here. The student is referred for further information to Heather's Treatise on *Drawing and Measuring Instruments**.

On closing the sector we find on each face of it, pairs of lines radiating from the centre of the circular metal hinge. These lines are known as the "sectoral lines" one line of each pair on either leg of the sector.

On one side we find three scales marked S, T, T. S is a line of *sines* graduated to 80°; T, T are lines of *tangents*, one extending from 0° to 45°, and the other computed to a less radius, extending from 45° to 75°.

On the other face of the instrument we find the sectoral lines marked L, S, C, Pol.

L is a line of *lines*, divided on each leg into 100 equal parts; S is a line of *secants* divided to 75°; C is a line of *chords* divided to 60°; Pol. is a line of *polygons* divided from 4 to 12.

It will be observed that these various scales are contained within three parallel straight lines, and that the extremity of the innermost of these lines is marked by a small brass nail. All distances from the centre are to be *measured along the line thus distinguished,* as it is the only one of the three which passes through the centre.

DEF. A distance taken along a sectoral line beginning at the centre is called a LATERAL DISTANCE.

DEF. A distance taken from any point in one leg to the corresponding point in the other leg is called a TRANSVERSE DISTANCE.

35. USE OF THE SECTORAL LINES.

THE LINE OF LINES (L).

These lines are divided into 10 equal parts, and each primary

---

* p. 98. Vol. 168 in Weale's Rudimentary Series.

part contains 10 equal subdivisions, hence the line is divided into 100 equal parts.

Example (1). Divide a straight line 2 inches long, into 10 equal parts.

Take 2 inches as a *lateral distance* between the points of the compass, and then open the sector until the *transverse distance* 10—10 is 2 inches, then it follows (by Euclid VI. 4) that the transverse distances 9—9, 8—8, 7—7, &c., are ·9, ·8, ·7, &c. of the two inch base 10—10, and these lengths being laid off from the same extremity of the line to be divided will give the points of section required.

Example (2). Find ·73 of 2 inches.

Take the lateral distance 2 inches, open the sector until the transverse distance 100—100 = 2 inches. Then take the transverse distance 73—73, and this will evidently be ·73 of 2 inches.

Example (3). On a map, a distance of 1·6 inches represents 80 miles, complete the scale to 100.

Make a transverse distance 80—80 equal to 1·6 inches; then take the transverse distance 100—100. Divide this into 10 equal parts, and subdivide the first primary into 10 parts.

Example (4). Find $\frac{7}{9}$ of 1·7 inches.

Make the transverse distance 9—9 equal to 1·7 inches; then take the transverse distance 7—7, which will be the length required.

Example (5). Find $\frac{72}{85}$ of ·8 of 4 inches.

Make the transverse distance 10—10 equal to 4 inches, and take the transverse distance 8—8. Then make the transverse distance 85—85 equal to this distance 8—8, and take 72—72.

Example (6). Find $\frac{35}{78}$ of $\frac{7}{9}$ of ·87 of $\frac{2}{3}$ of 5·3 inches.

Example (7). On a map, 7·3 miles = 4·2 inches, complete the scale to 100 miles.

Example (8). Construct a scale of equal parts, on which 7 inches will represent $1\frac{1}{4}$ inches.

36. LINE OF CHORDS (C).

This is divided on each leg into 6 parts of 10 degrees each and the degrees are further divided into two parts of 30' each. These lines are graduated by the method already explained (p. 38).

These double lines of chords on the sector have this advantage over the single scale on the Protractor, that in the former we may use *any radius* less than the widest transverse distance 60—60 on the sector, whereas in the latter the radius must be equal to the length 0—60 on the scale.

Example (1). To lay off an angle of 50°.

Draw a straight line AB, and describe an arc BC. Open the sector until the transverse distance 60—60 is equal to AB; then without changing the sector, take off the transverse distance 50—50, transfer this distance to BD on arc BC, and the angle $BAD = 50°$.

Example (2). To lay off an angle of 107°.

Divide by 2, then lay off chord of $53\frac{1}{2}°$ found as in Example (1) twice.

Examples for Exercise. Lay off angles of

38°; 49° 30'; 56° 45'; 79° 15'; 117° 20'; 145° 40'; 169°.

From a circle whose radius is 4 inches, cut off a segment capable of containing an angle of 43°, and find the length of the chord. *Result.* 2·95 inches.

37. LINE OF POLYGONS (Pol.).

This scale lies nearest to the inner edges of the instrument. Its divisions are marked 4, 5, 6, ...12. It has been so graduated that the transverse distance 5—5 will be contained as a chord *five* times in the circle whose radius is the transverse distance 6—6; the transverse distance 7—7, will be contained *seven* times, &c.

Example. In a given circle whose radius is 4 inches, inscribe a regular pentagon.

Make the transverse distance 6—6 equal to 4 inches. Take the transverse distance 5—5, and step it round the circle 5 times. Join the adjacent points, and a regular pentagon will be inscribed.

44   MARINE SURVEYING.   [CHAP.

Note. These constructions require much precision, otherwise the polygons will not "close" exactly. The circle ought to be drawn with a very fine pencil or steel bow, the dots marking the points ought to be placed in the middle of the circumference.

If it is required to describe a regular pentagon on a given straight line, we must proceed thus:—Make the transverse distance equal to the given straight line, and then take off the transverse distance 6—6. From each extremity of the given line as centre describe a circle with this radius thus found. The intersection of these arcs will be the centre of the circle, in which the polygon may be inscribed, whose sides are equal to the given straight line.

Examples for Exercise.

(1) On a given straight line $1\tfrac{1}{2}$ inches long, describe a regular heptagon.

(2) On a given line 1·4 inches long, construct a regular pentagon.

(3) On a line 1·5 inches long, construct a regular decagon.

38. We will conclude this part of our Chapter on Scales, &c. by drawing attention to the instruments known as Marquois Scales. These are named after their Inventor.

They consist of two box-wood rulers, 12 inches long, and a right-angled triangle of the same material. The triangle has its hypotenuse *three times the length of the shorter side.*

Near the edges on each face of the rulers will be found a pair of scales, the one nearer to the edge is called the "artificial scale," and that immediately behind it is known as the "natural scale."

These scales are divided into lengths of 10 units, the artificial scale along its entire length, but only the first division of the natural scale is so divided. In all cases however *one division of* 10 *units on the artificial scale is equal to three divisions of* 10 *units each on the natural scale,* and hence three natural units are equal to one artificial unit.

The zero of each scale is at its middle point. The numbering of the artificial scale proceeds right and left from the zero; and that of the natural scale from the extreme left of the scale to the extreme right.

Beneath the zero of each scale stands a number (15, 20, 40, 50, or 60) which denotes the number of units into which an inch is divided.

39. Use of these Scales.

Let it be required to draw two parallel straight lines at a distance of $\frac{2}{60}$ of an inch from each other. Draw one straight line, and place the longer side of the triangle coincident therewith: then place the ruler with *the zero of the scale* 60 in exact coincidence with the index at the middle point of the hypotenuse of the triangle. Keep the ruler firm and slide the triangle along *two artificial units*, and draw the line parallel to the former; these lines will be separated by $\frac{2}{60}$ inch. The triangle must be moved to the *left* if the line is to be drawn *above* the former, and to the *right* if it is to be drawn *below*.

Examples for Exercise.

(1) Draw 2 parallel lines 4 inches long and separated by $\frac{1}{60}$ inch.

(2) Draw 3 parallel straight lines 3 inches long, the first two $\frac{7}{60}$ of an inch from each other, and the third $\frac{11}{60}$ from the second.

(3) Draw 9 parallel straight lines 6 inches long and separated by $\frac{2}{15}$ inch from each other.

(4) Draw 5 parallel straight lines $2\frac{1}{2}$ inches long and distant from each other $\frac{9}{15}$ inch.

(5) Draw 4 parallel straight lines, the first and last pair distant $\frac{1}{8}$ inch, and the other pair $\frac{2}{3}$ inch.

40. We shall introduce here two important Problems in Geometrical Drawing, which are of frequent occurrence, and are particularly required in chart drawing. These problems are

(I.) To erect a perpendicular to a given straight line at any required point.

(II.) To divide a given straight line into a required number of equal parts.

41. I. To erect a perpendicular to a given straight line.

Let AB be a straight line, and it is required to erect a perpendicular at the point B.

Assume *any* point O above the line AB (the point ought not to be very near the line), and describe a circle with O as centre

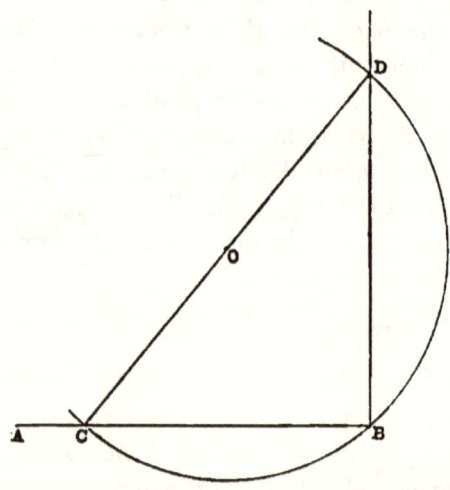

and OB as radius: let the circle cut the line in the point C. Join CO, and produce the line CO to cut the circle in the point D, then if D and B be joined, the line BD is perpendicular to AB.

By Euclid III. 31, the angle in a semicircle is a right angle, ∴ the angle $CBD = 90°$.

Note. In practice it is necessary to draw only two small portions of the circle CBD, viz. the part which cuts the given line at C and the part about D, so that the line CO produced will intersect it.

42. The student is strongly recommended to use a pricking-point in all these constructions, as by the use of this instrument

II.] THE CONSTRUCTION AND USE OF SCALES. 47

the points of intersection are more clearly defined than by any other method.

It may also be remarked that the longer the radius of the circle is taken, the less error will be produced in the required perpendicular through any slight mistake in the exact position of the point D. This is so self-evident that nothing further need be said.

43. II. TO DIVIDE A STRAIGHT LINE INTO A NUMBER OF EQUAL PARTS.

(i.) BY TRIAL.

Bisect the whole line, then bisect each part, &c.

But this method enables us only to divide into n parts, where n is some multiple $(2.4.8...)$ of 2. It is tedious, and unless great care is exercised, not accurate.

(ii.) BY PROPORTIONAL COMPASSES.

Set the index to the number required on the scale marked "Lines." Take the length of the line to be divided between the two points of the compass; reverse the instrument, and step off the distances.

In chart drawing this is certainly the most expeditious method if the student is in possession of this instrument.

(iii.) BY A SCALE OF EQUAL PARTS.

Let AB be a line to be divided, suppose, into 13 equal parts.

Draw BC making any angle with AB.

Take *any* scale of equal parts (for short lines the scale on the edge of the protractor seems well adapted) and placing its zero at A cause its 13th division to coincide with the line BC. Then with a pricking-point carefully make a mark at $A_1 A_2 A_3$......, and either by means of parallel rulers, or by a Marquois triangle and ruler, draw lines parallel to BC. These lines will divide the line AB into 13 equal parts.

(iv.) BY MARQUOIS SCALES.

Select one of the natural scales, a certain number of units on which will be equal to the line which is to be divided; then placing the triangle with its long side over one end of the line, place the zero of the scale selected coincident with the index, and keeping the ruler quite steady slide the triangle along the artificial scale making the index coincide with the necessary strokes: as each stroke is reached make a dot on the given line with a pricking-point;—e.g. divide a line one inch long into 6 parts. Select the scale 60. Make the index coincide with every 10th stroke on the artificial scale.

CHAPTER III.

LAYING OFF ANGLES.

44. An angle may be protracted by any of the following methods.

(1) *By semicircular Protractor.*

This instrument is graduated to degrees, and hence by estimation we can draw an angle within $\frac{1}{2}$ or $\frac{1}{3}$ of a degree.

(2) *By rectangular Protractor.*

This is made of various sizes. The usual size in cases of instruments is 6 inches long, and is graduated to degrees. We can estimate to within $\frac{1}{3}$ or $\frac{1}{4}$ of a degree, and in larger sizes to within $\frac{1}{6}$ of a degree.

(3) *By the circular Protractor.*

By the aid of the attached verniers we can protract to *minutes*.

(4) *By the scale of chords on the Protractor, or the lines of chords on the Sector.*

By estimation to within $\frac{1}{2}$, $\frac{1}{3}$, or $\frac{1}{4}$ degree.

(5) *By Construction,* we can lay off *accurately* an angle of any given dimensions as follows.

Note. This last method is called "PROJECTION BY CHORDS."

45. Let AOB represent any angle a. Take OA equal to any given radius. Describe the arc AB. Bisect OA in C, and describe the arc CD. Draw the chords AB, CD. Bisect the arc AB, then the line joining the centre O with this point of bisection will bisect the chords in E and F respectively and also the angle a.

50 MARINE SURVEYING. [CHAP.

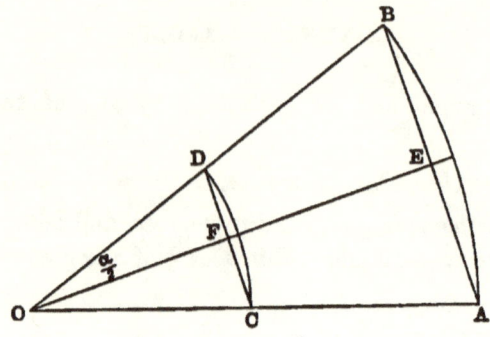

By Euclid VI. 4, $\dfrac{CD}{AB} = \dfrac{OC}{OA} = \tfrac{1}{2}$, ∴ $CD = \tfrac{1}{2}AB = EB$.

But $\dfrac{EB}{OB} = \sin\dfrac{a}{2}$, ∴ $\dfrac{CD}{OB} = \sin\dfrac{a}{2}$,

∴ $CD = OB \sin\dfrac{a}{2}$.

Hence the practical rule:

"Multiply the natural sine of half the given angle by twice the length of the given radius, and then with centre O and a line (called the *effective radius*) equal to the given radius, describe a circle. Lay off from a scale of equal parts the chord CD equal to the length thus found; join OD, and the angle DOC will be the required angle."

46. But how is the *natural* sine computed?

In two ways: (1) Take out the tabular log sine, and subtract 10, then take out the natural number corresponding to the result.

e.g. Compute the natural sine of 48° 26'.

```
Tab. log sine 48° 26' = 9·874008
    Subtract            10
                        1·874008
                          3960     = ·7481
                            48     ........8
                            46
                            20     ......... 3
                            17
                            30     ........... 5
```

∴ nat. sine 48° 26' = ·7481835.

III.] LAYING OFF ANGLES. 51

(2) Since $\text{vers}(90 + a) = 1 - \cos(90 + a) = 1 + \sin a$,

$$\therefore \sin a = \text{vers}(90 + a) - 1$$

Hence the natural sine is most expeditiously computed as follows: Increase the given angle by 90° and take out the tabular versine of this increased angle. Divide the tabular versine by 1000000 (because the tabular versines are a million times the natural versines) by placing a decimal point after the left figure; subtract 1, and the result will be the natural sine of the given angle.

e.g. Compute the natural sine of 48° 26′,

```
                48° 26′
       Add       90
                ──────
                138 26
    Tab. vers 138° 26′ =  1748184
    Divided by 1000000 = 1·748184
    Subtract                   1
                          ──────
    Natural sine 48° 26′ =  ·748184
```

the same result as above.

Example. Protract an angle of 32° 18′ accurately with radius 4 inches.

We evidently require the natural sine of 16° 9′,

```
                16° 9′
       Add       90
                ──────
                106 9
    Natural versine = 1·278153
    ∴ natural sine =  ·278153
    Twice radius =           8
                     ─────────
    Chord required = 2·225224
```

With radius 4 inches long describe a circle. Lay off the chord of 2·225 inches or 4·45 half-inches. Join the centre and the other extremity of the chord, and the angle between the two radii is 32° 18′ as required.

Examples for Exercise.

(1) Protract an angle of 37° 40' accurately with radius equal to 5 inches. (Oct. 1875.)

(2) Protract an angle of 48° 30' with radius 5 inches. (Nov. 1876.)

(3) Protract an angle of 74° 40' with radius 5 inches. (June, 1877.)

(4) Erect a perpendicular to a straight line using the scale of chords, and also by means of the compass and ruler only, marking and describing the construction in each case. (May, 1878.)

(5) Divide a straight line 5·8 inches in length into 10 equal parts; at one end erect a perpendicular, and from the other end draw a line which shall make with the given line an angle of 65° 30'. (Nov. 1878.)

(6) By means of the scale of chords, protract the following angles: 39°; 65°; 118°. Upon what theory does the application of this scale to angular measurement depend? (Aug. 1879.)

(7) Protract an angle of 117° 25' accurately with radius of 4·5 inches.

CHAPTER IV.

FIXING POSITIONS BY ANGLES.

47. WE must in the first place investigate a method of describing a segment of a circle capable of containing an angle of a given size. In Euc. III. 33 we have a geometrical method of construction, but in practice it is much more easily accomplished.

The following geometrical theorems must be known:

(1) The angles at the base of an isosceles triangle are equal.

(2) The three angles of a triangle are together equal to two right angles.

(3) The angle at the centre of a circle is double the angle at the circumference standing on an equal arc.

(4) The opposite angles of a quadrilateral inscribed in a circle are together equal to two right angles.

48. There are three cases to be considered, viz. when the angle is right, acute, or obtuse.

By Euc. III. 31, the angle in a semicircle is right; the angle in a segment greater than a semicircle is acute; and in a segment less than a semicircle the angle is obtuse.

CASE I. *When the given angle is right.*
On the given line as base describe a semicircle, and any angle in this is 90°.

CASE II. *When the given angle is acute.*

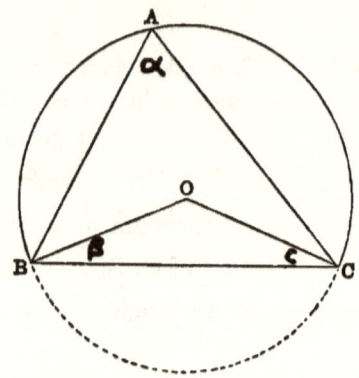

Let $BAC = a$, and BC the given line;

$$\angle OBC = \angle OCB;$$

$$\therefore 2OBC + BOC = 2 \text{ right angles.}$$

But $\qquad BOC = 2BAC$ (Euc. III. 20);

$$\therefore 2OBC + 2a = 180;$$

$$\therefore OBC + a = 90;$$

$$\therefore OBC = 90 - a = OCB.$$

Also, a being less than 90°, the segment BAC is greater than a semicircle, and therefore the centre of the circle and the angle a are on the same side of BC.

Hence the rule: "At each extremity of the given line BC lay off an angle CBO, BCO equal to the *complement* of the given angle; then with centre O and distance OB describe a circle BAC; BAC shall be a segment capable of containing an angle equal to a."

CASE III. *When the given angle is obtuse.*

Let $BAC = a$, and let D be any point in the conjugate segment BDC.

FIXING POSITIONS BY ANGLES.

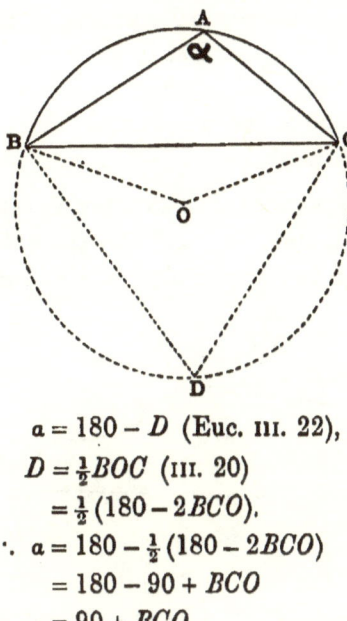

$$a = 180 - D \text{ (Euc. III. 22)},$$
$$D = \tfrac{1}{2} BOC \text{ (III. 20)}$$
$$= \tfrac{1}{2}(180 - 2BCO).$$
$$\therefore a = 180 - \tfrac{1}{2}(180 - 2BCO)$$
$$= 180 - 90 + BCO$$
$$= 90 + BCO.$$
$$\therefore BCO = a - 90.$$

Also, BAC being obtuse, the segment BAC is less than a semi-circle, and therefore the centre of the circle is on the side of BC remote from a.

Hence the rule: "At B and C the extremities of the given line lay off angles CBO, BCO equal to the excess of a above 90°. With centre O and distance OB, or OC describe a circle, and the smaller segment shall be capable of containing an angle equal to a."

Examples for Exercise.

(1) On a straight line 4·35 inches long describe a segment capable of containing an angle of 39°, of 78°, and of 122°.

(2) On a straight line $2\tfrac{1}{2}$ inches long describe a segment capable of containing an angle of 18° 12′, of 37° 25′, and 129° 57′.

(3) A and B are two points 5500 yards apart; construct on the line AB segments which shall contain angles of 72° 30′; 141°; 155° 30′. Scale $\tfrac{1}{2}$ inch = 1000 yards.

49. There are two principal methods of fixing a position in the survey of a harbour: the first is known as the "straight line and one angle method," and is chiefly used in running lines of soundings; the second is the well-known "Three-point Problem," with its various modifications.

50. (1) THE STRAIGHT LINE AND ONE ANGLE METHOD.

Suppose it is required to "fix" the point D situated on the line passing through two known objects A and B.

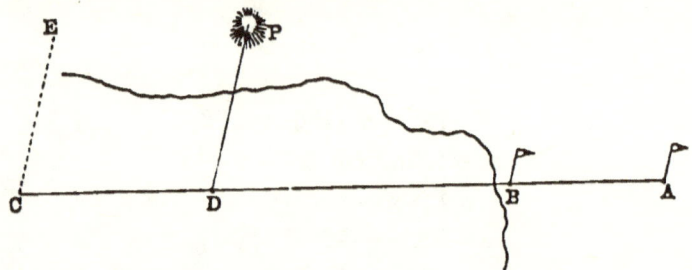

Observe the angle BDP between B and some well-defined object P, which must not be too far away. At any point C in the line AB produced make the angle ACE equal to the angle just observed. Then a line through P, a known point in the survey, parallel to EC, will cut the line at D.

If a segment capable of containing an angle equal to the observed angle be described on BP as chord or base, it will intersect the line at the point D.

Note. In using this method it is well to observe, if possible, an angle on both sides to check the accuracy of the work.

51. (2) THE THREE-POINT PROBLEM.

This consists in observing at a station the angle between two points A and B, also between two points B and C, and then describing on the lines AB, BC respectively segments of circles capable of containing angles equal to those just observed. The segments will intersect at the stations required.

52. There are SIX CASES of this problem.

Let the station at which the angles are observed be denoted by the letter P.

IV.]   FIXING POSITIONS BY ANGLES.   57

Case I. *When P is outside the triangle formed by joining the three points A, B, C, and the central object is on the side of the line joining the other two remote from P.*

1st method. Let $APB = a$, $CPB = \beta$.

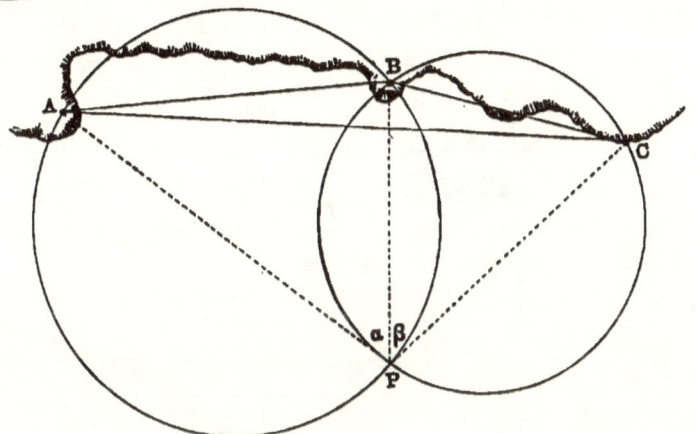

By the methods already explained describe on AB, BC, segments containing angles a, β respectively, these segments will intersect in P, the position of observation.

When the point P is fixed by the intersection of two circles, the method is called the "TWO CIRCLE METHOD" of projection.

2nd method. A, B, C are the three points. Make the angle CAD = observed angle β, and the angle ACD = observed angle a.

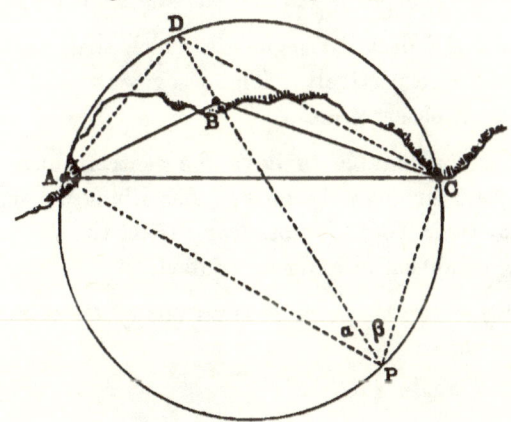

Describe a circle about the triangle ACD; join DB, and produce it to cut the circle in P.

Then, by Euc. III. 21, the angle CPD = angle $CAD = \beta$ as observed, and also the angle APD = angle $ACD = \alpha$ as observed. Hence P is the position of observation.

Note 1. When the points D and B are very close together, the 1st method is advisable.

Note 2. When this 2nd method is practicable it is known as the "STRAIGHT LINE AND CIRCLE METHOD" of projection.

CASE II. *When P lies outside the triangle, but the central object observed lies on the same side as P.*

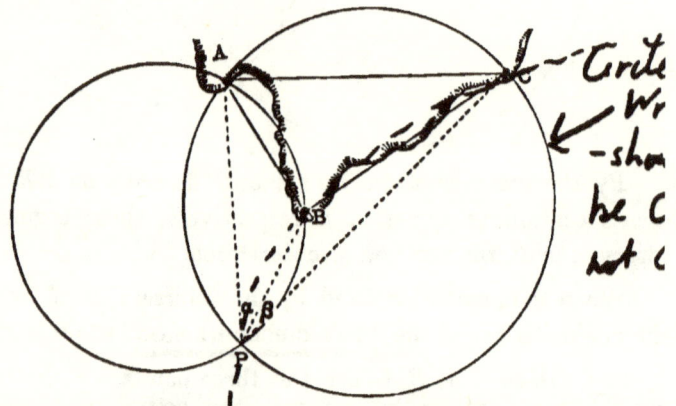

Observe the angle $APB = \alpha$, and the angle $CPB = \beta$.

On BC and AB describe segments which shall contain angles equal to β and α respectively. These segments will intersect in P the position of observation.

Note. It is undesirable to have the central object near and the other objects far away, because on describing the segments we shall find that their "cut" is not clear, in fact the two circles may approach the condition of external contact.

CASE III. *When the point P is on one of the sides of the triangle ABC produced.*

Observe the angle $APB = \alpha$.

The points A, B, C being known points in the survey, the angle BAC is known.

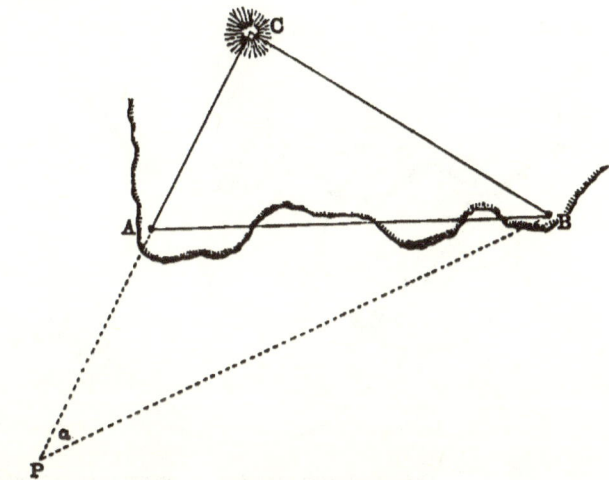

Then, Euc. I. 32, $CAB = a + ABP$;
∴ $ABP = CAB - a.$

Hence ABP is known.

At point B lay off this angle thus found, and BP will intersect CA produced in the position of observation.

Note. Comparing this case with the straight line and one angle method (p. 56) we see that the latter is really a special case of the three-point problem.

Case IV. *When the point of observation is on one of the sides of the triangle ABC.*

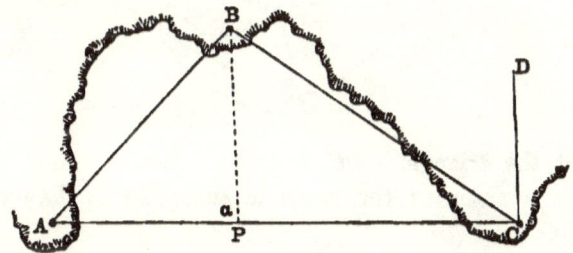

Observe the angle $APB = a$.

At C make the angle ACD equal to a. Then a line drawn through B parallel to CD will intersect AC in the point of observation P.

MARINE SURVEYING.   [CHAP.

Case V. *When the three objects A, B, C are in the same straight line.*

1st method. Observe the angle $APB = \alpha$, and the angle $CPB = \beta$.

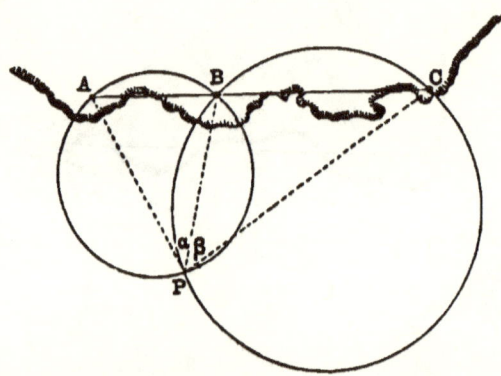

Then on AB, CB respectively describe segments capable of containing angles α, β respectively. These segments will intersect in P the position of observation.

2nd method. Lay off $ACD = \alpha$, and $CAD = \beta$.

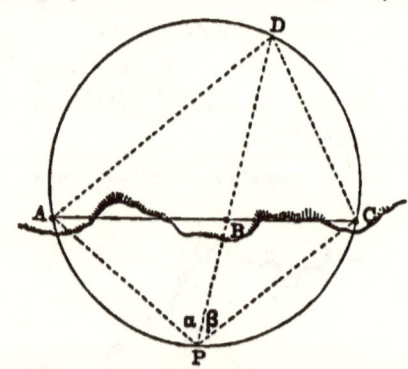

About the triangle ADC describe a circle. Join DB, and produce it to intersect the circle in the point P, the position of observation.

Join PA, PC. Then
$$APD = ACD = \alpha,$$
and $$CPD = CAD = \beta.$$
Hence P is the position of observation.

FIXING POSITIONS BY ANGLES.

CASE VI. *When the point P is within the triangle formed by joining the given objects.*

1st method. Observe the angle $APB = \alpha$, and the angle $BPC = \beta$. On AB, BC respectively, describe segments of circles

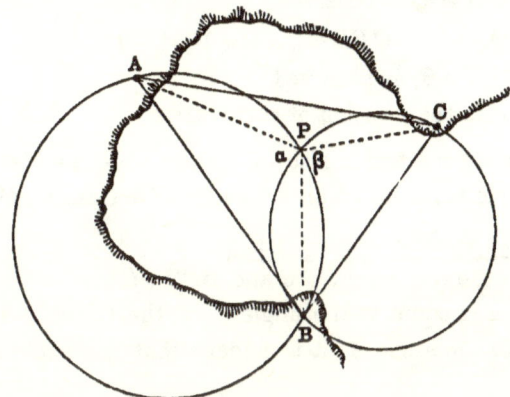

containing angles α and β respectively. These segments will intersect in P, the position of observation.

2nd method. Observe the angles $APB = \alpha$, $BPC = \beta$, and $CPA = \gamma$ (as a check).

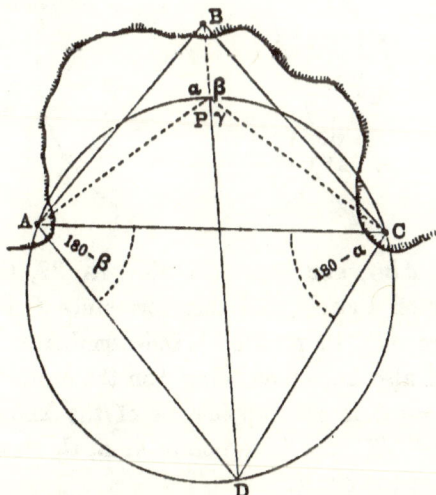

At C make $ACD = 180° - \alpha$, and at A make $CAD = 180° - \beta$. These lines will intersect in D.

About the triangle ADC describe a circle. Join BD, and this line will intersect the circle in P the position of observation.

By Euc. III. 21,

$$\text{angle } APD = \text{angle } ACD = 180° - a;$$

therefore $BPA = a$, as observed.

Similarly, $\qquad CPD = CAD = 180° - \beta;$

therefore $BPC = \beta$, as observed.

Hence P is the position of observation.

This second method is also an example of the "straight line and one circle method" of projection. *Vide* Case I., Method 2.

53. *The Indeterminate Case.*

Observe the angle $APB = a$ and $BPC = \beta$.

Now if a is equal to the angle C of the triangle ABC, and if β is equal to the angle A, it is evident that the angle APC is the

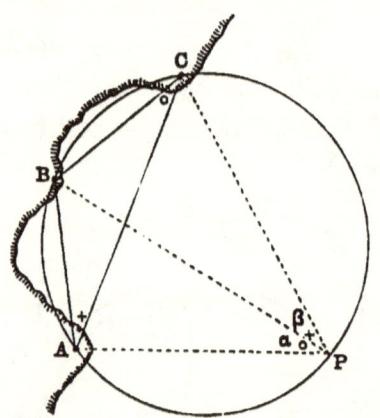

supplement of ABC, and hence, by Euc. III. 22, the four points A, B, C, P are on a circle, and therefore since P may be at any point of the arc APC its position is indeterminate.

This case is also known as being "on the circle."

Whenever $a + \beta$ is the supplement of the known angle ABC of the triangle ABC, the four points A, B, C, P will lie on the circumference of the circle.

If however no other points are available for observation, then the compass must be resorted to, and one or more bearings taken.

Note. In the last figure we have the two segments coincident, and therefore the centres of the two circles coincide. We see then that *if the centres of the two circles are very near each other* the segments will not give a clear "cut," and therefore the point of intersection will not be sharply defined. The best "cuts" of course are those which most nearly approach a right angle

In using the three-point problem some little experience is necessary in the selection of suitable objects. The observed angles should be as near 90° as possible, ought not to differ much from each other; the objects ought to be nearly equidistant, not too far distant, and finally no angle less than 25° or 30° ought to be admitted, unless the central object is very distant, when a small angle between it and one of the others is considered good.

54. The best relative positions of the points in fixing a position by means of the Three-point Problem are:

(1) If the three objects are in the same straight line.

(2) If the middle object and the position of observation are on the same side of the line which joins the other two objects.

(3) If the position of observation is within the triangle formed by the three objects.

Under these circumstances the Indeterminate Case, or its approximations can never occur.

If it be required to fix a point by an observed angle of 70° between A and B, and an angle of 85° between B and C, we use the notation "$A\,70°\,B\,85°\,C$."

Hence "hut 56° tree 68° rock" means that the angle observed between the hut and tree was 56° and between the tree and rock was 68°. The hut, tree, and rock are therefore the "three points," or objects.

55. The following methods of fixing a position by angles observed between three points *without actually describing the segments* which contain those angles seem to be worthy of notice by the reader:—

Let the observed angles be $A\,a°\,B\,\beta°\,C$.

Suppose the circles to have been described (Case II., above) and the point P has been fixed by their intersection, then we obtain the following analysis. Draw BD, BD' the diameters of the circles. Join CD, AD', and draw PA, PB, and PC.

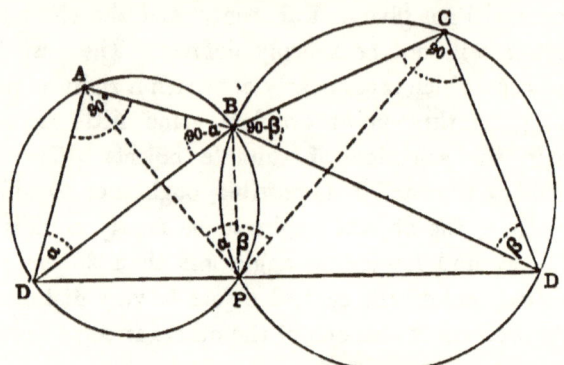

By Euc. III. 31, the angle $BCD = 90°$, and by Euc. III. 21, the angle $CDB = CPB =$ the observed angle $\beta°$. Hence the angle $CBD = 90° - \beta$. Similarly we have the angle $ABD' = 90 - \alpha$. By III. 31, the angles BPD and BPD' are right angles, and therefore DD' is a straight line (I. 14).

Example. A bears from B WNW. one mile, C bears from B

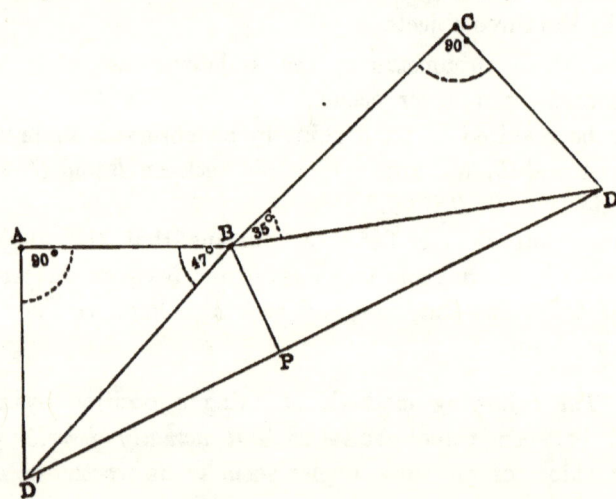

NE^{b}E. 1·5 mile. From P we observe A 43° B 55° C. Fix the position of P. Scale 1 inch to a mile.

IV.] FIXING POSITIONS BY ANGLES. 65

Draw BA, BC according to the given scale. Make $CBD = 35°$ (the complement of 55°) and draw CD at right angles to CB. Make $ABD' = 47°$ (the complement of 43°), and draw AD' at right angles to AB. Join DD' and let fall BP perpendicular to DD'. P will be the position of observation.

The following seems to be a satisfactory method of protraction without drawing the circles.

Let the notation be $A\,a°\,B\,\beta°\,C$.

Let P be fixed by the intersection of the two circles in the

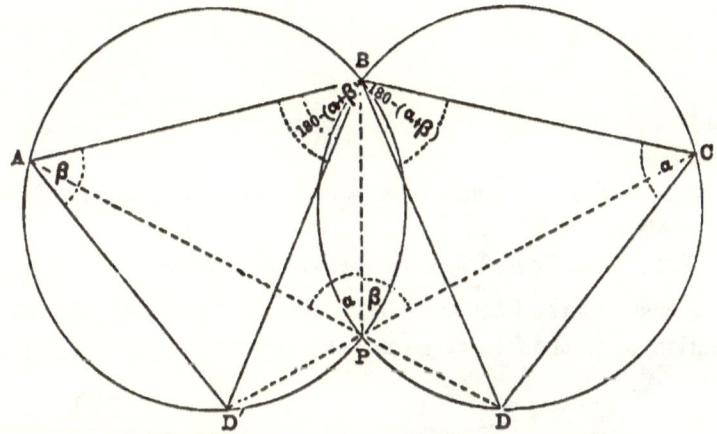

usual way. Join PA, PB, and PC. Produce AP and CP to meet the circles again in D and D' respectively.

Join DB, DC, $D'A$, $D'B$.

Then we have by Euc. III. 21, the angle $CBD = CPD = 180 - (\alpha + \beta) =$ supplement of the sum of the observed angles.

Again $CBPD$ is a quadrilateral inscribed in a circle, therefore by Euc. III. 22, we have the angle $BCD = BPA = \alpha$. Similarly we get $ABD' = 180 - (\alpha + \beta)$ and $BAD' = \beta$.

We may therefore fix the position of P as follows.

Let the notation be $A\,80°\,B\,44°\,C$.

The supplement of the sum of the observed angles $= 56°$.

Make $CBD = 56°$ and $BCD = 80$ (the angle subtended by the points A, B). Let the lines meet in D. Make $ABD' = 56°$ and

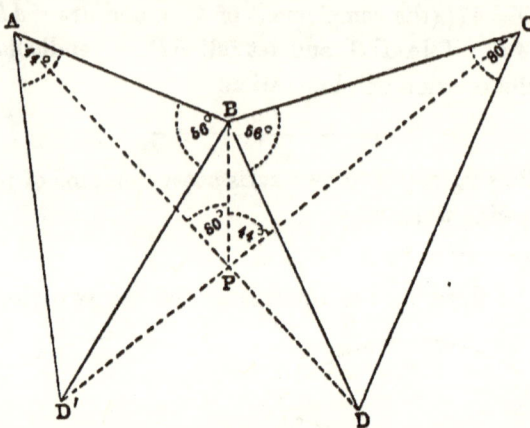

$BAD' = 44$ (the angle subtended by the points B, C). Let the lines meet in D'.

Join CD', AD. These lines will intersect in P the position of observation.

Note. The lines AD, CD' may be called *Lines of Position*.

Sometimes the lines must be produced to meet in the observer's position, as in the following example.

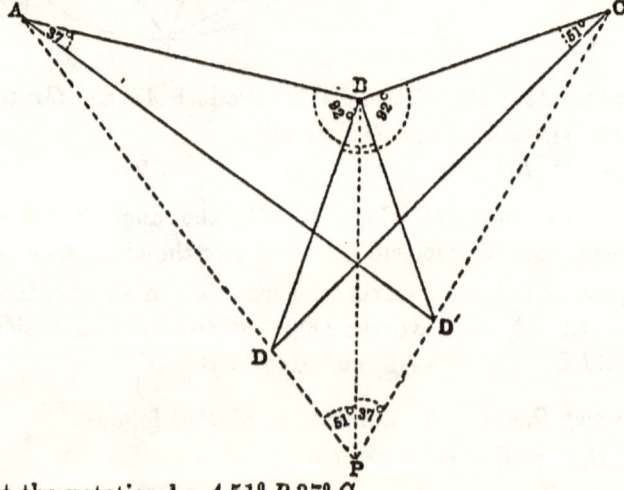

Let the notation be $A\, 51°\, B\, 37°\, C$.

FIXING POSITIONS BY ANGLES.

The supplement of the sum of the observed angles $= 92°$. Make $CBD = 92°$ and $BCD = 51°$. Make $ABD' = 92°$ and $BAD' = 37°$. Join AD, CD' and produce them to meet in P, the observer's position.

The reader may draw the general figure, and prove that this will be the case.

Note. This method of fixing a position is known as the "STRAIGHT LINE METHOD."

The following example of the straight line and circle method is very good.

Let the notation be $A a° B \beta° C$.

Lay off $ACD = a°$, and $CAD = \beta°$. Describe a circle about the triangle ACD. Join B, or B_1, or B_2, or B_3 (the middle object)

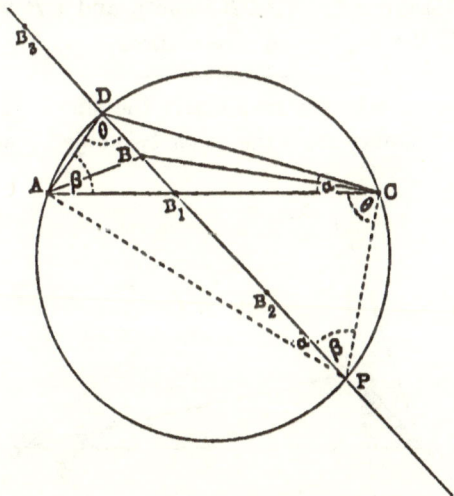

with D, and produce the line indefinitely to cut the circle again in P. Then it is evident by Euclid III. 21, that $APD = ACD = a$, and $CPD = CAD = \beta$. And also $ACP = ADP$, and $CAP = CDP$.

Hence we have the following method of fixing the position of P without describing the circle.

Let A, B, C be the points, and let the notation be $A\ 30°\ B\ 62°\ C$.

68 MARINE SURVEYING. [CHAP.

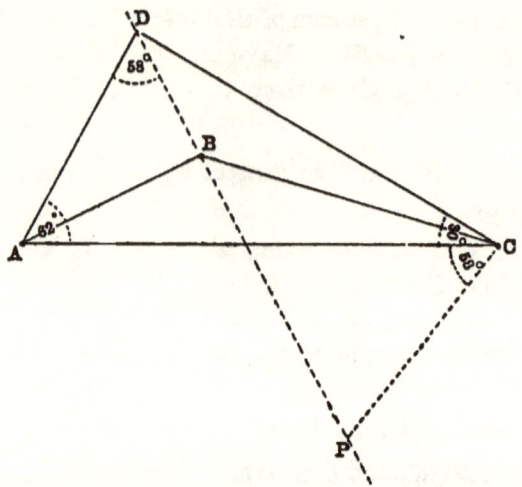

Make $ACD = 30°$, and $CAD = 62°$. Join DB and produce it indefinitely. Make $ACP = ADB$ (= 58°), and CP will meet DB produced in P, the position of observation.

Finally, let us take the case where the middle (B) object is at a considerable distance from the other two objects, and where one

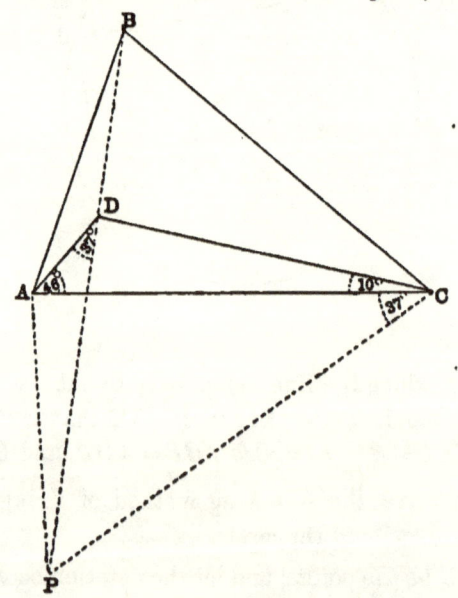

of the angles observed is very small. In this case the radii are so long that it is impracticable to describe the circle.

Let the notation be $A\ 10°\ B\ 46°\ C$.

Make $CAD = 46°$, $ACD = 10°$.

Join BD, and produce it indefinitely. Lay off $ACP = ADP$, which will be found to be 37°, and CP will intersect BD in P, the position of observation.

Before attempting the general exercises at the end of this Chapter the student will find it useful to project each of the following examples by the One-Circle Method, the Two-Circle Method, and the Straight-Line Method.

(1) When both the observed angles are *less than* 90°
Let the Notation be $A\ 43°\ B\ 70°\ C$.
$AB = 4000$ feet; $BC = 3600$ feet; $CA = 5800$ feet.
Scale ½ in. = 1000 feet.

(2) When both the observed angles are *greater than* 90°
Let the Notation be $A\ 110°\ B\ 125°\ C$.
$AB = 5200$ feet; $BC = 3900$ feet; $CA = 5500$ feet.
Scale ½ in. = 1000 feet.

(3) When one of the observed angles is *greater than* 90°, and the other angle is *less than* 90°
Let the Notation be $A\ 115°\ B\ 74°\ C$.
$AB = 6200$ feet; $BC = 5000$ feet; $CA = 7200$ feet.
Scale ½ in. = 1000 feet.

Danger Angle.

56. Def. The DANGER ANGLE is the angle subtended at a shoal or other hidden danger by two well-defined permanent objects.

By Euclid III. 21, *all* angles in the same segment of a circle are equal; hence if on observing the angle between these two known permanent objects, the angle is found to be *equal* to that laid down on the chart as observed from the shoal, it follows that the ship is *somewhere on* the segment and therefore in danger; if the angle is *greater than* the chart angle the ship is *inside* the segment, and therefore probably in danger; but if the angle is *less than* the given angle the ship is *outside* the segment and in a position of safety, and will continue so as long as the observed angle is less than the given chart angle.

The annexed diagram will make this clear.

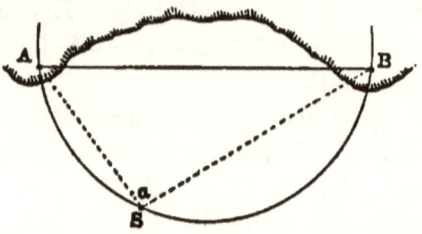

A and B are the permanent objects, and S the position of a shoal.

The angle $ASB = a$ is observed, and on AB a segment is described capable of containing an angle equal to a. This angle a is called the "danger angle," the segment is known as the "danger segment," and it is evident that as long as the observed angle taken on board between A and B is *less than* the angle a, the position of the ship is outside the danger segment.

57. The following problems are added as being sometimes of use when instruments of the usual kind are not available.

Problem I. To find the length of a line accessible only at the ends.

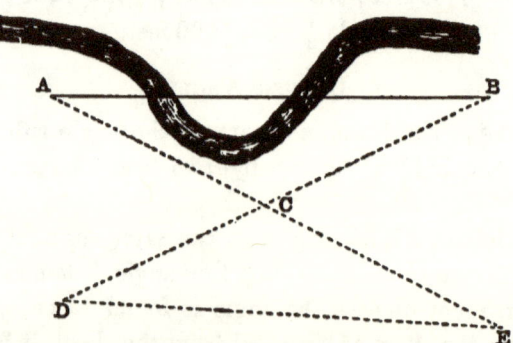

Let the distance AB be required; select a point C, and set up a mark. Pace BC, and then pace CD equal to BC and in the same direction. Similarly pace $CE = AC$. Then $DE = AB$. (Euc. I. 4.)

FIXING POSITIONS BY ANGLES.

Problem II. To find the distance of an inaccessible object by the rhombus method.

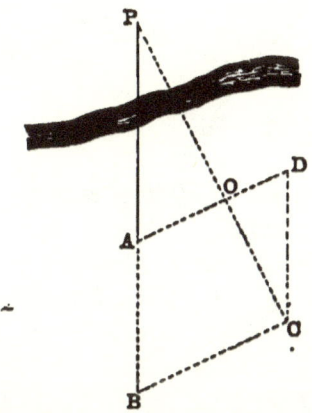

Let P be the point on the opposite side of a river, and let the distance BP be required.

Take a known length of string and tying a knot at its middle, fix the knot at B, and fasten one end of the string to a peg at A in the direction of P and the other end to a peg situated in any convenient direction C: then removing the knot to the position D, the string being kept taut, mark D, and place a peg at O where OP cuts AD.

The triangles ODC, CBP are similar,

$$\therefore \text{ by Euclid vi. 4,} \quad \frac{BP}{BC} = \frac{DC}{DO}, \quad \therefore BP = \frac{BC \cdot CD}{DO},$$

and these quantities on the right-hand side of the equation are known, hence BP may be found.

E.g. suppose the side of the rhombus was 100 feet, and the distance DO was 11 feet, then

$$BP = \frac{100 \times 100}{11} = \frac{10000}{11} = 909 \text{ feet, approximately.}$$

Problem III. To find the distance of the point where two lines intersect in a river or lake.

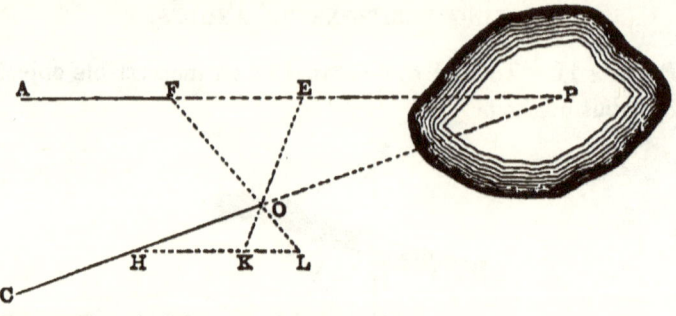

Let AF and CO meet in P, it is required to determine the length of the line FP, or OP.

Join FO and make $OL = \frac{1}{2}OF$. Join any other point E in AP with O, and make $OK = \frac{1}{2}OE$. Join LK and produce it to cut OC in H. Then the triangles LOH, FOP are similar (Euc. VI. 6; VI. 4),

$$\therefore PO : HO = FO : OL = 2 : 1.$$

HO can be measured, and hence OP is found.

EXAMINATION.

(1) Mention the two principal methods of fixing a position in marine surveying.

(2) What conditions ought to be looked for in selecting the objects in the three-point problem?

(3) You are passing along a coast, and take with a sextant two angles between three known points: how would you fix the ship's position by the station pointer?

(4) Specify the propositions in Euclid's Third Book on which the theory of the three-point problem depends.

(5) Specify the various possible cases of this well-known problem.

(6) Mention those cases in which no ambiguity is possible.

(7) Define the danger angle. Draw a figure to illustrate your answer.

(8) What do you understand by the expression "on the circle"?

(9) How would you proceed to find the approximate width of a river whose banks are straight and parallel, if unprovided with the usual instruments?

(10) Suppose the banks are irregular how would you proceed?

(11) Two objects at the ends of a given line are accessible but the line itself cannot be measured directly, how would you set about obtaining the approximate distance?

(12) Define and illustrate the methods of projection known as "the two circle method," "the straight line and one angle method," "the straight line and one circle method," and the "straight line method."

(13) How may the ambiguity implied in question 8 be remedied?

EXAMPLES FOR EXERCISE.

(1) Project the following on the scale of 1 inch = a mile
bluff 67° jetty 106° hut.

From bluff, jetty bore N. 7° W. 2·9 miles and hut bore N. 37°30′ E. 5·4 miles. Determine by projection the bearing and distance of the point of observation from the jetty. (Dec. 1874.)

(2) From mosque, palm bears N. 82° E. (true) 1·9 miles, and a shoal S. 63° E. (true) 1·45 miles. Protract on a scale of 2 inches = a mile.

Assuming it to be unsafe, standing in on a northerly course with the shoal ahead, to approach nearer than 3 cables, state the "danger angle" which the shore-points would then subtend.

(Feb. 1877.)

(3) From flag, spire bears N. 15° E. (true) 1·2 miles, and mound bears S. 36° E. 1·33 miles. Project on a scale of 3 inches = a mile. Project also the following line of soundings, reduction being 3 feet.

At $4^h 10^m$, spire 39°30′, flag 30°20′, mound; 20 feet; 19 ft.; 18 ft.; 18 ft.

At $4^h 24^m$, spire 46° 40′, flag 34° 50′, mound; 17 ft.

(April, 1877.)

(4) From flag, mound bears N. 34 E. (true) 6990 feet, and the angle mound (on the left) to tree is 58° 30′.

At mound, tree 66° 45′ flag,
At tree, flag 54° 45′ mound.

Project the position of the tree by calculated sides; the scale being 3 inches to a mile of 6082 feet. (May, 1877.)

(5) From A on jibboom at end of a frigate the angle between B on starboard quarter and battery flagstaff was 80° 10′, and simultaneously the angle observed at B between A and the flagstaff was 79° 5′. The horizontal distance between A and midship gun was 100 feet, and the distance between A and B was 300 feet. Protract on a scale of 2 inches to 300 feet, and ascertain the distance of the flagstaff from the gun. (Nov. 1877.)

(6) Two torpedoes are submerged in line on a bearing E^{b}N. and 200 yards apart. The western torpedo bears N. 600 yards from a lighthouse, the eastern bears NW. 400 yards from a fort. The coast line (of sand) runs straight between the lighthouse and fort. The bearings are *true*. Protract the figure on a scale of 1 inch = 400 yards, and shew the *true* course a vessel from the westward must steer in order to pass in mid-channel between the torpedoes and the sandy coast. (March, 1878.)

(7) Stations A, B, C are in a line on a NW^{b}N. bearing. From B, D bears N. 17° 15′ E. distant one mile. At D the angle between C and $B = 51°$, between C and $A = 77° 30′$. Protract these positions on a scale of 3 inches = a mile.

Fix (1).   Fix (2).
$B \phi D 86° A$, $D 50° B 35° A$.

Place these "fixes" on the plan, and find the distance between C and Fix (2). (April, 1878.)

(8) At A the angle between B and $C = 120°$. At B the angle between C and $A = 30°$. The distance from A to $B = 1·5$ miles. At D, the angles C 63° A 65° B were observed. Protract on a scale of 4 inches = a mile of 6050 feet. Place D on the plan, and calculate the natural scale. (May, 1878.)

IV.]   FIXING POSITIONS BY ANGLES.   75

(9) From fort, cliff bears SSE. 2½ miles, and point bears SW. 2 miles. The bearings are true. At wreck, point 75° fort 89° cliff. Protract on a scale of 1 inch = a mile, and mark the position of the wreck by two methods.   (Sept. 1878.)

(10) At hill, lighthouse 90° cliff; at cliff, hill 60° lighthouse. The hill bears north from cliff distant 1·5 miles.

A ship to the southward of cliff and lighthouse observes a *rock awash* between her and the shore, and in passing takes the following bearings:

Rock φ Lighthouse N. 14° E.; rock φ hill N. 50° W.; rock φ cliff N. 80° W.

Bearings are *magnetic* throughout, variation = 11° E. Protract the above on the scale of 2 inches = a mile 6050 feet, and place the rock awash correctly on the plan.   (Oct. 1878.)

(11) Three buoys, A, B, C, moored on the outskirts of a sandbank are equidistant from each other 3000 yards. A bears from B N. 33 E. (*Mag.*); C is to the eastward of A and B.

At A.   At B.
———→ sandbank 8° 30' B. A 7° 30' ←——— sandbank.
←——— sandbank 47° 30' B. A 51° 15' ———→ sandbank.

At C.
B 8° 45' ←——— sandbank.
B 49° 30' ———→ sandbank.

Variation = 17° W. Protract on a scale of 1·8 inches = 2000 yards.

Rule the True and Mag. Meridians through B, and dot in the sandbank.   (Nov. 1878.)

(12) From ship, a rock bore N. 8° E., and after steering W^{b}S. 1·5 miles it bore N. 63° E. Protract on scale of 3 inches = a mile, and give the distances of the ship from the rock at the time of each observation. Bearings magnetic. Variation = 11° W.
(March, 1879.)

76   MARINE SURVEYING.   [CHAP.

(13) From summit of hill A 300 feet above the sea, the point B bears S. 33 W. distant 1·2 miles, and the theodolite angle between B and 3 rocks awash in line to the eastward was 28°; the angles of depression being respectively 5° 42' 40"; 4° 17' 20"; and 3° 26' 00. Bearings magnetic. Variation 12° E. Protract on a scale of 2·5 inches = a mile of 6063 feet, and give the magnetic bearings of each rock from the point B.   (May, 1879.)

(14) The top of a lighthouse 180 feet above the sea, subtends a vertical angle to the water line, from A of 9°, bearing N. 43° E. from B of 11°, bearing N. 53° W. A buoy to the southward lies equidistant from A and B 300 yards. If the light be obscured within a horizontal distance of 380 yards, how far from this limit is the position of the buoy? Var. = 3° E. Bearings magnetic. Protract on a scale of 0·9 in. = 100 yards.   (Aug. 1879.)

(15) A bears from B W^{b}S. distant $\frac{3}{4}$ mile.
       D ,,   ,, B S^{b}E.   ,, $\frac{1}{2}$,,.
       C ,,   ,, A S^{b}W.   ,, $\frac{6}{10}$,,
Var. = 10° W. Bearings magnetic.

At E., A 95° B 58° 30' D. Protract on a scale of 2·2 inches = a mile. Fix the position of E, and from it give the true bearings of A, C, D. Rule the true and magnetic meridians through A.

(Sept. 1879.)

(16) From ship, magnetic bearings of

A = N. 31 E. distant 1·1 mile.
B = S 73 E.   ,,   1·4 ,,
C = S 29 W.   ,,   0·8 ,,
D = N 24 W.   ,,   0·7 ,,

Var. = 5° E. Protract on a scale of 1·2 inches = a mile of 6060 feet, and give the two bearings and distances of these points from each other.   (Dec. 1879.)

(17) When sailing along a coast, a headland was observed to bear NE. $\frac{3}{4}$ N. (true); having run E. $\frac{1}{2}$ N. (true) 14 miles, the headland bore W^{b}N. $\frac{1}{4}$ N. (true): required its distance from the ship at each observation, and protract its position on a scale of 0·5 inches = a nautical mile.   (May, 1880.)

IV.]   FIXING POSITIONS BY ANGLES.   77

(18) From house, the following true bearings and distances were observed and measured:—Tree N. 58° E. 2½ miles; church E. 5° S. 4½ miles; light vessel E. 22° S. 3 miles. Protract the above on a scale of 1·5 inches = a mile, and plot the position of the following soundings:

House 108° Tree 97° Church (2 feet, mud).
House 112° Church ϕ Lighthouse (3¼ fathoms, coral).
House 41° Light vessel 44° Church (7 fms., ooze).

(Aug. 1880.)

(19) From church the following *true* bearings and distances were observed and measured; Tree N. 75° E. 5 miles; Windmill S. 81° E. 6 miles; Lightship S. 68° E. 3 miles. Protract on a scale of 1·1 inches = a mile, and mark the position of the following shoals:

Church 137° 30' Tree 77° 30' Windmill (3¼ fms., hard sand).
Church 61° 30' Light ship 38° 50' Windmill (5 fms., mud and ooze).

(20) From church the following true bearings and distances were observed and measured:—Lighthouse N. 83° E. 7¼ miles; Tree S. 82° E. 2¾ miles.

From a vessel the following angles between these objects were observed. Church 46° Tree 48° Lighthouse.

How must the vessel steer so as to pass one mile south of the lighthouse? Protract on a scale of 1 inch = a mile. (Nov. 1880.)

(21) From Lighthouse, Windmill bears S. 65° E. (true) 1½ miles, and Church bears S. 85° E. (true) 3 miles.

A vessel southward of windmill observes a rock awash between her and the shore, and takes the following bearings (true) while passing;

Rock ϕ Lighthouse N. 50° W.; Rock ϕ Windmill N 29° W.;
Rock ϕ Church N. 40° E.

Protract the position of the Rock on a scale of 2 inches = a mile.

(Apr. 1881.)

(22) From Nubble, Fort bears N. 15° W (true) 3900 feet, and the angle

Fort (on the left) to Quoin is 70° 00'.
At Fort   Quoin 62° 30' Nubble.
At Quoin   Nubble 47° 30' Fort.

Project the position of Quoin by calculated sides. Scale 6 in. = a mile of 6080 feet. (May 1876.)

(23) From Coastguard, Mound bore N. 77° W. (true) 0·45 of a mile, and Mill bore N. 88° E. 0·56 of a mile; the following stations were taken to fix a shoal on which the sea breaks too heavily to risk the boat near. Mound 60° 0' C. G. 47° 0' mill.

ϕ
Centre of shoal.
Mound 55·0 C. G. 57° 30' mill.
ϕ
Centre of shoal.

Project the positions on a scale of 5 inches = a mile; giving the centre of the shoal. (Sept. 1876.)

---

(24) The mean of a set of observations taken at Pile △, Lat. 52° 0' N., was as follows. Zero, Steeple 360° 0' Mag. Bearing S. 32° W. $5^h 16^m$ p.m. A.T. 85·00 ϕ. Sun's decl. 20° 30' N.
Required true bearing of Steeple from Pile, and the variation.
(June, 1877.)

(25) The mean of a set of observations taken at Theodolite △ X, in Lat. 53° 14' N., was as follows:—

Sun's decl. = 16° 38' N.   Semi = 15' 50".
Zero-Camp △ 360° 0'   Mag. N. 45° E.

$6^h 0^m$ a.m. A. T. 37° 25'   ☉

Required true bearing of the Camp from X, and the Variation.
(April, 1876.)

(26) The mean of a set of observations taken at Cairn △, in Lat. 50° N., was as follows:—Sun's decl. = 12° 30' N.

Zero, Wedge △   360° 0'   Mag. N. 30° 0' E
$6^h 0^m$ a.m. A. T.   40° 3'   ϕ

IV.] FIXING POSITIONS BY ANGLES. 79

Required the true bearing of the wedge △ from the Cairn △, and the Variation. (Oct. 1876.)

(27) At Theodolite △ X, in. Lat. 54° 3′ N., the following observations were made:

Zero, Lighthouse (Southward of X)......360° 0′.

First set with sun's *lower* and *right* limbs.

$3^h\ 58^m$ p. m. 35° 11′ alt.......104° 19′.
35° 6′ ,, ...104° 26′.
35° 1′ ,, ...104° 33′.

Second set with sun's *upper* and *left* limbs.

$4^h\ 0^m$ p. m. 35° 30′ alt.......104° 4′.
35° 26′ ,, ...104° 11′.
35° 21′ ,, ...104° 18′.

Sun's decl. = 19° 36′ 6′ N.

Required the true bearing of the Lighthouse from X. (Aug. 1875.)

Answers.

Note. The distances in these results are for the most part given in *inches*, as actually taken off from the paper. The student can then ascertain the correctness of his work without difficulty.

(1) S. 71° E. 2·4 miles.

(2) 63°.

(3) Length of line of soundings = ·83 in. = ·28 mile.
  1st sounding S. 46½ W. from spire.

(4) Mound to tree 3·65 in.; flag to tree 3·92in.

(5) 5·55 inches.

(6) N. 75 E.

(7) Fix 1 to B = 2·15 in.; fix 2 to C = 1·90 in.

(8) Nat. Scale = $\frac{1}{18150}$. DB = 5·65 in.; DC = 5·75 in.

(9) Wreck bears from point S. 15½ W. 2 miles.

(10) Distance from lighthouse = 3.85 in.; from hill = 5·72 in.

(11) If x, y, z be the points of the bank nearest to A, B, C respectively, then xy = 1·37 in.; yz = 1·32; zx = 1·47 in.

80   MARINE SURVEYING.

(12) From 1st position 1·43 in. = ·715 miles.
    2nd „   5·13 in. = 2·565 „ .

(13) The rocks are 3000, 4000, 5000 feet respectively from A, and bear N. 42 E.; N. 52$\frac{1}{2}$ E.; N. 65 E.

(14) 17$\frac{2}{3}$ yards. Distance from A to B = 4·29 inches.

(15) EB = 1·23 miles; A = N. 64° W.; C = S. 48$\frac{1}{2}$ W.; D due East.

(16) From A, B bears S. 24 E. 1·87'.
     B, C ...... S. 85$\frac{1}{2}$ W. 2·12.
     C, D ...... N. 9$\frac{3}{4}$ E. 1·61.
     D, A ...... N. 75$\frac{3}{4}$ E. 0·60.

(17) Distance from 1st position = 2·73 miles; from 2nd position = 5·5 miles.

(18) No. 1, sounding from house = 3·0 in.; no. 2, from house = 3·5 in.; no. 3, from light vessel = 2·03; from church = 3·86 in.

(19) No. 1, shoal from church = 4·25 in.; from tree = 1·60.
    No. 2, shoal from light vessel = 1·95 in.; from windmill = 4·75.

(20) Ship from church 3·47 in. Course = S. 57$\frac{1}{2}$ E.

(21) Rock to lighthouse = 3·42 in.
    Rock to windmill = 2·16 in.

(22) Quoin to nubble = 4·8 in.
    „ fort = 5·0 in.

(23) Centre of shoal S. 13 W. 1·44 in.

(24) Steeple bears S. 9° 39' W.
    T. B. of sun N. 85° 21' W. Var. = 12° 21' W.

(25) T. B. of sun N. 79° 52' E. Camp bears from T. N 42° 11' E. Variation 2° 49' W.

(26) N. 41° 50' E. Var. = 11° 50' E.

(27) S. 27° 16' 50" E.

CHAPTER V.

CHARTS AND CHART DRAWING.

58. A MAP is a representation on a plane of a large portion of the earth's surface.

A CHART is a representation on a plane of a portion of the earth's surface of large or small extent with special reference to the requirements of the seaman.

A PLAN is a representation of a very limited extent of the earth's surface, drawn on a plane without reference to the latitude or longitude of the positions; it has respect only to the relative position and distances of points, for which purpose a scale of distances is annexed. Thus we have plans of harbours, anchorages, &c.

Note. Plans are frequently placed in the corners of charts for more special use, such plans will give more detailed information than could be conveyed on the chart itself.

59. DEF. A representation of a figure on a surface formed by the intersection of that surface by lines drawn from the observer's eye to every visible point of the figure is called a *Projection.*

The plane on which this representation takes place is known as the "*Primitive Plane,*" or simply the "*Primitive,*" or the "*Plane of Projection.*"

Projections are of two kinds, *Natural* and *Artificial.*

A *Natural* Projection is simply a perspective delineation of any object on the Primitive.

An *Artificial* Projection is not a perspective representation.

A Projection of the Sphere is thus a representation of the surface of the sphere on a plane.

The Natural Projections of the sphere are representations of the surface of the sphere on a plane as seen from a certain position of the observer.

60. The principal Natural Projections are the ORTHOGRAPHIC, the STEREOGRAPHIC, and the CENTRAL or GNOMONIC.

DEF. The Orthographic can be explained as follows: Suppose the eye placed at an infinite distance, then all lines drawn from the eye to the sphere will be parallel; accordingly, if perpendiculars be let fall from every point on the surface of a hemisphere on its diametral plane as Primitive, the representation will be such as it would appear to the eye at an infinite distance. In this case, however, only the central portions are correctly delineated, whereas the portions near the edges are unduly crowded and distorted. This Projection is therefore only of use for the representation of small portions of the Earth's surface.

61. DEF. In the Stereographic Projection the eye must be conceived as situated at one extremity of a diameter of the sphere, and as viewing the concave surface of the sphere through the diametral plane as a Primitive. E.g. If the eye is placed at the South Pole and views the Northern hemisphere through the plane of the Equator, then every point on the surface will appear delineated on the plane of the Equator.

In the former Projection the points as they recede from the centre are crowded together, in this Projection on the contrary their projected dimensions seem to be somewhat enlarged.

Two properties of this Stereographic Projection make it important: (1) All circles on the hemisphere are represented by circles on the Projection, and (2) all small triangles on the surface of the sphere are represented by triangles similar to them in the projection. This valuable property insures a general similarity of appearance in the map to the reality, and enables a single hemisphere to be represented in a single map.

62. DEF. In the Central, or Gnomonic Projection, the eye is supposed to be situated at the centre of the sphere, and the

Primitive is a tangent plane to the sphere. Since the plane of every great circle passes through the eye, in this Projection every great circle will be represented by a straight line on the Primitive, and hence charts on this Projection are well adapted for Great Circle Sailing*. The circumpolar regions are usually represented on this Projection. It is evident that a complete hemisphere cannot be thus represented, because the great circle which bounds it is on a level with the eye, and therefore parallel to the Primitive plane. The Maps in ordinary Atlases are constructed by this Projection.

63. It may also be observed that if in the Gnomonic Projection the Primitive plane touch the Earth at any point between the Equator and the Pole, the meridians will be projected as straight lines which will meet at the projection of the Pole. The angle contained between two meridians is called their *Convergency*. Now from the definition of a True Bearing it follows also that if two points differ in Latitude the T. B. of the point in the lower Latitude exceeds the reverse T. B. by the value of the convergency of the meridians of the two places.

64. MERCATOR'S PROJECTION.

This is known as an Artificial Projection, and is named after the inventor, who was born in Flanders early in the 16th century. The mathematical principles of its construction seem to have been first enunciated by Edward Wright, of Cambridge, at the end of the same century.

Suppose the eye situated at the centre of the earth, regarded as a hollow crystal sphere. On the surface we may conceive the equator, parallels, and meridians to be painted black, and a hollow cylinder, of infinite length, painted white on the inside, and touching the earth at the equator. Now it is evident that the black lines on the sphere will be projected on to the white background, and also that the intervals between the parallels of latitude will increase as the eye ranges from the equator towards the pole: at the pole the line of sight coincides with the axis of the earth, and

* Godfray's "Chart to facilitate the Practice of Great Circle Sailing" is an example; the Primitive being a tangent plane at the South Pole.

only meets the circumscribing cylinder at infinity. In other words the meridional parts of 90° = infinity.

Yet another method may be suggested of getting a clear conception of a Mercator's chart. Suppose the earth as before to be circumscribed by a cylinder painted white, only now conceive the earth to be made of some material capable of infinite expansion. If all the features on the surface are recently painted, then as the expansion can take place only *in the cylinder*, we can understand that the features coming into contact with the cylinder will leave traces in paint on the white ground, but the parts about the pole will be at an enormous distance from the equator, and the pole itself at infinity.

Thus then in Mercator's Projection the various places on the earth are correctly represented as to their *form*, but the scale varies greatly. In the polar regions, for example, the scale is extravagantly enlarged; we may have a small island in a high latitude, *seemingly* as large as India.

If the cylinder on which the paint from the expanding earth has been transferred be opened or unrolled, we shall have represented a Mercator's chart of the world.

65. Plane Chart.

Since a small portion of the surface of the globe may be considered as a plane without sensible error, it is evident that charts of coasts, harbours, islands, anchorages, rivers, &c., may be constructed without reference to their special latitude and longitude. The object of such a chart is to obtain bearings and distances, and these can be easily obtained from such plane charts.

In places near the equator the degrees of latitude and longitude may be considered as equal*, and therefore a plane chart of regions of low latitude may be constructed of considerable extent. If a plane chart were constructed for regions of large extent in high latitudes, then no direction is correctly shewn except due N. and S., E. and W., and no true distance except on the meridian.

* To realise this, let the student consult his Table of Meridional parts for the first 10 or 15 degrees of latitude, and he will notice how little they increase.

These plane charts of great extent are no longer constructed, Mercator's charts having completely superseded them.

66. In drawing plans of harbours, &c., it may be noted that the *Nautical Mile* used is the *Minute of Latitude in*. As the earth is an oblate spheroid the *degree* of latitude increases as we advance from the equator towards the pole, and hence its sixtieth part, or the nautical mile, also increases*.

67. The two great advantages of Mercator's chart are, (1) All Rhumb lines between two places on the chart are *straight lines*, and (2) the angle at which this line cuts the parallel meridians is the *Course*.

The great disadvantage is that although all features retain their *correct form*, yet the *scale* for different latitudes varies much, and in very high latitudes it becomes so much exaggerated, that a portion of land within the Arctic circle may appear two or three times the size of an equal portion within the Tropics.

METHOD OF CONSTRUCTING A MERCATOR'S CHART.

68. It seems advisable to notice the following point in the first place for the sake of the younger students. The meridional parts for $49° = 3382·08$ and for $50° = 3474·47$, the difference being $92·39$. Now what does this difference mean? We may explain it thus. On the principle of construction employed in Mercator's projection, the degrees of latitude increase as we go from the Equator towards the Pole. A degree of the Equator $= 60$ miles; and the above difference of $92·39$ miles signifies that a degree of latitude extending from 49 to 50 contains $92·39$ equatorial miles.

69. The construction of the chart may be divided into the following steps:

(1) To find the meridional parts.

(2) To compute the lengths of the meridional difference of latitude on the given scale.

* The length of a degree of latitude increases as follows:
  In lat. $12°$ the length is 362956 feet.
  ,,......52 ,,............364951 ......
  ,,......66 ,,............365744 :......

(3) To draw the longitude line, and to divide it according to the given scale.

(4) To erect the perpendiculars at the ends of this longitude line, and to test the accuracy of the work.

(5) To set off on the meridian thus drawn, the lengths as found in (2).

(6) To draw the upper parallel.

(7) To rule in the meridians and parallels.

These steps constitute the construction of the frame of the chart.

(8) To draw outside the frame another fine line about $\frac{1}{10}$ or $\frac{1}{15}$ inch distant from, and parallel to it.

(9) To divide the degrees each into six equal parts of 10'.

(10) To lay down the latitude and longitude from.

(11) To lay down the true courses and distances.

(12) To find the latitude and longitude in.

(13) To find the bearing and distances to or from any known position.

(14) To finish off the chart according to individual taste.

70. We shall illustrate the principal parts of the above in discussing the following chart.

Construct a Mercator's chart on a scale of 1·38 inches to a degree of longitude, extending from latitude 48°N. to 51°N., and from 20°W. to 26°W.

To compute the mer. parts and the lengths of the mer. diff. latitudes according to the scale, we arrange the work thus:

$$51° = 3568·81 \qquad 50° = 3474·47 \qquad 49° = 3382·08$$
$$50 = 3474·47 \qquad 49 = 3382·08 \qquad 48 = 3291·53$$
$$\text{diff.} = \overline{\quad 94·34 \quad} \qquad \text{diff.} = \overline{\quad 92·39 \quad} \qquad \text{diff.} = \overline{\quad 90·55 \quad}$$

We know from above what these differences imply.

Now, by hypothesis, 60 miles at the Equator are represented by 1·38 inches; we have therefore to find what number of inches will represent 94·34 miles, 92·39 miles, and 90·55 miles respectively.

We have evidently to work three Simple Proportion sums. Let us take the first.

As 60 miles : 1·38 inches :: 94·34 miles : x inches;

$\therefore\ 60x = 94\cdot34 \times 1\cdot38;$

$\therefore\ x = \dfrac{94\cdot34 \times 1\cdot38}{60}$ inches;

or, $\dfrac{94\cdot34 \times 1\cdot38}{60} \times 2 =$ value of x in $\frac{1}{2}$ inches;

$\therefore\ \dfrac{94\cdot34 \times 1\cdot38}{30} =$ value of x in $\frac{1}{2}$ inches.

Hence the well-known Rule: "Multiply the difference of the meridional parts by the scale, and divide the result by 30; the result will be the length to be taken from the $\frac{1}{2}$-inch diagonal scale on the Protractor."

We may notice that the length for 94·34 is greater than that for 92·39, and this in turn is greater than that for 90·55, and this is as it should be, the degrees increasing as we get farther from the Equator.

Before we proceed further, the accuracy of these different computations ought to be tested as follows :—

Find the mer. diff. between the two extreme latitudes and compute its length on the scale by the usual method. This length ought evidently to be equal to the sum of the separate parts already found.

We thus know the size of our chart, and can therefore place it suitably on the drawing paper.

71. ERECTION OF THE PERPENDICULARS.

We recommend the method of drawing a perpendicular explained in the Second Chapter. The pencil in the compasses ought to have a very fine point, and the points of intersection ought to be marked with the pricking-point. So much depends on these lines being carefully drawn that the student ought to bestow every attention on this part of the work.

The accuracy of the construction may be tested as follows. Open the compass to any convenient length (about 5 or 6 inches), and measure off equal lengths along the perpendiculars from the longitude line; then the distance between these points thus found ought to be equal to the length of the longitude line.

The fine line drawn outside the frame is best drawn by the aid of the Marquois Scales, otherwise the eye is the sole judge.

72. The division of the degrees into equal parts seems on the whole to be most expeditiously performed by means of the proportional compasses. For other methods of dividing a line, the student is referred to the Chapter on Scales. We recommend the graduation to be carried all round the chart, and not merely along the longitude line and one meridian.

In ruling all lines care must be taken that the ruling pen (which ought to be of the very best make, and ought to be guarded from all rough treatment) is well supplied with ink*, that the ruler is held firmly in its place, that the pen is drawn evenly along the edge, and that the motion proceed from the shoulder joint and not from the wrist. Before any line is drawn on the chart with the pen, the pen ought to be tested *on paper of the same kind* as that on which the chart is being constructed.

73. To lay down the Latitude and Longitude of a point on a Chart.

Place a ruler along the latitude of the place found from the graduated meridians, and draw a very fine pencil line about $\frac{1}{4}$ inch in length under the meridian of the place as judged by the eye. Then lay down the ruler to coincide with the longitude as marked on the top and bottom parallels, and draw a similar line, or make a point on the line drawn before; this will mark the position of the place required.

The converse process will enable us to find the latitude and longitude of a place on the chart.

---

* We can confidently recommend Rowney's Fluid Indian Ink, sold in small bottles at one shilling.

To lay down a Course on a Chart.

Place the index of the Protractor over the point from which the ship sailed, so that the lower edge may be exactly parallel to the nearest parallel of latitude, and then prick off the course from the graduated edge.

To lay down a Distance on a Chart.

We know the *direction* it will take; estimate with the eye the approximate middle point of the line, and take the distance from the graduated meridian, one half the distance above and the other half below the approximate latitude of this middle point. Lay the edge of a ruler through the points, marking the place and the course, and lay off the distance now found.

The converse process will enable us to find the course and distance between two points*.

74. We have noticed that the two following points present difficulties to some of the younger students, we shall therefore notice them before concluding this part of our subject.

Suppose the chart is to extend from 53°N. to 55° 40′N., and from 29° 10′ W. to 32° 40′ W. Where ought the parallel to be drawn?

We would compute the meridional parts for the following pairs:

55° 40′	55° 00′	54′ 00′
55° 00′	54° 00′	53° 00′

The space between 55° and 55° 40′ is only 40 miles. The parallels might be drawn at 54° and 55°, but the space between 55° and 55° 40′ must be divided into only *four* equal parts, the spaces between the other parallels being divided into *six* parts.

The meridians may be drawn at every ½ degree, e.g. 29° 10′, 29° 40′, 30° 10′, &c., or at every *even degree*, the first two meridians being in this case only 10 miles apart, and the last two being 40 miles apart. In such cases however it seems best to space

---

* The student is advised to read Raper's remarks on the "Properties of Certain Projections" in his *Navigation*, p. 127.

the chart so that the appearance of the meridians and parallels may be most pleasing to the eye. The *principle* is not in any way interfered with.

75. Suppose the scale is given for a *middle latitude*, how are we to proceed?

E. g. In the December Chart (1875) the scale was 2·13 inches to 1° of middle latitude, the chart to extend from 49° to 52° N. and longitude 9° to 16° W.

The scale for 1° longitude may be found in two ways.

First method. Find the middle latitude of the chart; in this case 50° 30′. This is the middle of the degree 50°—51°. Find the meridional difference of latitude, thus:

$$\begin{array}{r} \text{mer. parts for } 51° = 3568\cdot81 \\ \text{,,} \qquad\quad 50° = 3474\cdot47 \\ \hline \text{diff.} \quad = \quad 94\cdot34 \end{array}$$

Now we know from what has gone before that if 60 miles at the equator be represented by x inches, we can compute what number of inches will represent 94·34 miles: and conversely, if we have given the number of inches which represent 94·34 miles, we can find what number of inches will represent 60 miles at the equator, and this will be the required scale.

Hence as $\quad 94\cdot34 : 2\cdot13 :: 60 : x;$

$\therefore x = 1\cdot35$ inches, the required scale.

Second method. The degrees of latitude in Mercator's chart are increased in the ratio of the *secant of the latitude*, and hence

scale at the equator = scale at middle latitude

$\qquad\qquad\qquad\qquad\qquad \times$ cosine of the middle latitude;

$\therefore x = 2\cdot13 \times \cos 50° 30′,$

$$\begin{array}{r} 0\cdot328380 \\ 9\cdot803510 \\ \hline \log x = 0\cdot131890 \end{array}$$

$\therefore x = 1\cdot35$, as before.

Examples for Exercise.

(1) Mercator's chart. Scale 1·38 inches = 1° longitude.
From lat. 48° N. to 51° N. and long. 20° W. to 26° W.

  Courses E. ½N. Distances 148′. Variation 30° W.
    NNE.   73′.
    SSE.   123′.
    N^{b}E ½ E.   55′.

Ship sailed from lat. 48° 7′ N. and long. 25° 36′ W., and made ½-point leeway during the last course, the wind being east.

Lay down the true courses, and find the latitude and longitude in. (Feb. 1875.)

(2) Mercator's chart, 1·25 in. = 1° long.
From lat. 59° to 63° S. and long. 178° E. to 177° W.
Ship sailed from lat. 62° 50′ S. long. 178° 50′ E. as follows:
NE. 100′; W^{b}S. 80′; N. 125′; E^{b}N. ¾N. 25′; SW. ½W. 95′; E. ½S. 135′.
Variation 1¾ points E.
Lay down the true courses, and find lat. and long. in.
  (Aug. 1873.)

(3) Mercator's chart, 1·33 in. = 1° long.
From lat. 64° to 67° 30′ N. and long. 0° to 5° W.
Ship sailed from 64° 30′ N. and 3° 30′ W.
NE. ½ E. 145′; NW^{b}W. 66′; SSW. ½ W. 53′; SE^{b}E. ½ E. 83′; SW. 75′; NW. 60′.
Variation 2½ points W.
Lay down the true courses, and find the lat. and long. in.
  (Feb. 1874.)

(4) Mercator's chart, 5 in. = 1° long.
To extend from lat. 49° N. to 50° N. and from long. 8° W. to 9° 30′ W.

Ship sailed from 49° 50′ N. and 9° 20′ W.

Comp. co.	Dev.	Dist.	
S^{b}W.$\frac{1}{2}$W.	5° W.	30′.	
SE^{b}E.$\frac{3}{4}$E.	8° E.	35′.	Var. = 23° W.
N^{b}E.$\frac{1}{4}$E.	3° E.	35′.	

Find the latitude and longitude in. (March, 1881.)

(5) Mercator's chart, 1·65 in. = 1° long.
To extend from lat. 54° to 56° 30′ N. and from long. 14° to 20° W.
Ship sailed from 54° 27′ N. 14° 49′ W.,
N.$\frac{3}{4}$E. 87′; SW^{b}W. 103′; N^{b}W. 86′; N. 24′.
Variation = 28° W.
Observations then placed the ship in lat. 56° 12′ N. and long. 18° 13′ W.
Lay down the true courses, and find the direction and distance set by the current. (March, 1875.)

(6) Mercator's chart, 3·46 in. = 1° long.
To extend from lat. 57° to 58° N. and from long. 2° to 5° E.
Ship sailed from 57° 7′ N., 2° 11′ E. Corr. mag. E. 54′.
Variation = 20° W.
Observation then placed her in 57° 33′ N., 3° 27′ E. Find set and drift of the current. Speed of ship and the rate of current remaining the same, find the ship's course to reach lat. 57° 53′ N. and long. 4° 34′ E. (Oct. 1875; May 1876; Feb. 1877.)

(7) Mercator's chart, 3·6 in. = 1° long.
To extend from lat. 38° to 39° 30′ S. and from long. 77° to 80° E.
Ship left lat. 39° 12′ S. and long. 77° 9′ E.
E.$\frac{3}{4}$S. 82′; E^{b}N.$\frac{1}{4}$N. 28′.
Variation = 26° W.
An island in lat. 38° 13′ S., long. 79° 32′ E., was then seen bearing NE$\frac{1}{4}$N., and the course was altered to E^{b}S$\frac{1}{2}$S.; after making good 17·5 miles in that direction, the island bore NNW. The courses and bearings are cor. mag. Required the position of the ship, and the direction and distance set by the current.

(Aug. 1875.)

(8) Mercator's chart, 1·38 inches = 1° long.
To extend from lat. 60° to 62° 25' N. and from long. 19° to 25° W.
Find by Projection the cor. mag. bearing and distance of a port whose lat. is 62° 23' N. and long. 19° 57' W. from a ship in lat. 60° 11' N. and long. 24° 36' W.
Variation = 22° W.
State also the course that must be steered to reach the port, allowing for a set of N. 34°W. (true) 32 miles, and a deviation of 5° E. (April, 1875.)

(9) Mercator's chart, 1·65 inches = 1° long.
To extend from lat. 58° to 60° N. and from long. 0° to 6° E.
Ship sailed from lat. 58° 9' N. and long. 0° 22' E.
E. $\frac{1}{2}$ N. 75'; ENE. 61'; E^{b}N. 25'.
Variation = 22° W.
A lighthouse in lat. 59° 17' N. and long. 4° 54' E. then bore SSE$\frac{1}{2}$ E., and the angle to a point N. 23° E. (true) 20 miles from the lighthouse was 68°. The courses and bearings are cor. mag.
Required the direction and distance set by the current.
(Sept. 1875.)

(10) Mercator's chart, 2·13 in. = 1° of middle latitude.
To extend from lat. 49° to 52° N. and from long. 9° to 16° W.
Ship sailed from lat. 49° 25' N. and long. 15° 36' W.
E. $\frac{3}{4}$ N. 125'; E^{b}N. $\frac{1}{2}$ N. 98'.
Variation = 15° W.
A point of land in lat. 51° 40' N., long. 10° 4' W., then bore NE.; after running 30 miles farther on the same course E^{b}N. $\frac{1}{2}$ N., the point bore N^{b}W$\frac{1}{2}$ W.; the courses and bearings are cor. mag.
Required the position of the ship, and set and drift of the current, if any. (Dec. 1875.)

(11) Mercator's chart, 2·15 in. = 1° of middle latitude.
To extend from lat. 49° to 52° N. and long. 8° to 15° W.
Ship sailed from lat. 49° 20' N., long. 14° 15' W.
E. 100'; E^{b}N. 65'; E^{b}N. $\frac{3}{4}$ N. 40'.
Variation = 25° W.

A point of land in lat. 51° 40′ N., long. 9° 24′ W., then bore N^bE. (mag.), and the angle to a peak E.½S. (true) 23 miles from the point was 60°. Find the position of the ship and the current experienced. (Sept. 1876.)

(12) Mercator's chart, 2·22 in. = 1° middle latitude.
To extend from lat. 46° to 49° N., and long. 4° to 11° W.
Ship sailed from lat. 46° 30′ N., long. 10° 10′ W.
E. 120′; ENE. 106′.
Variation = 21° W.

Ushant Light in lat. 48° 28′ N., long. 5° 8′ W., then bore E^bS., and after running 32 miles farther on the same course ENE., the light was lost sight of bearing SW^bS. Courses and bearings are magnetic. Required the position of the ship, and the current experienced. (June, 1877.)

(13) Mercator's chart, 2·6 in. = 1° long.
To extend from lat. 64° to 65° N. and long. 16° to 20° W.
Ship sailed from lat. 64° 5′ N., long. 16° 11′ W.
NW^bW. 50′; W^bN.½N. 30′; W. 12′.
Variation = 5° E.

An islet in lat. 64° 58′ N., long. 19° 37′ W., was then seen bearing 4 points on the starboard bow, and after running 9 miles farther on the same course (west) the islet was exactly abeam. Find the position of the ship and the distance from the islet.

(April, 1876.)

(*Note.* A similar chart in every respect appeared in March, 1877.)

(14) Mercator's chart, 2·6 in. = 1° long.
To extend from lat. 49° to 50° 30′ N., long. 3° to 6° W.
Ship left lat. 49° 12′ N., long. 5° 40′ W.
ENE. 40′; E^bN.¼N. 41′.
Variation = 21° W.

The Eddystone in lat. 50° 11′ N., long. 4° 16′ W., then bore NNW.¾W. (mag.), and the angle to the Start which bears N. 85 E. (true) 25 miles from the Eddystone was 112°. Required the position of the ship, and the comp. course to be steered to pass 7 miles off the Start, allowing for a deviation of 5°·E.   (Oct. 1876.)

(*Note.* Same chart given in April, 1879, and very slightly altered in Oct. 1880.)

(15) Mercator's chart, 3·4 in. = 1° long.
To extend from lat. 36° to 37° 30′ S. and long. 70° to 73° E.
Ship left lat. 37° 20′ S., long. 70° 15′ E.
E. ½ N. 60′; E. ¼ S. 42′.
Variation = 23° W.
An island in lat. 36° 18′ S., long. 72° 30′ E., then bore NE^{b}E. ¼ E., and after running 12·5 miles farther on the same course E. ¼ S., the island bore NNE. ¼ E. The courses and bearings are magnetic. Required the position of the ship, and the current experienced.
(Nov. 1876.)

(16) Mercator's chart, 1·75 in. = 1° long.
To extend from lat. 36° to 40° S. and long. 20° to 24° W.
Ship left lat. 36° 20′ S., long. 20° 30′ W.
SW. 60′; W^{b}S. 100′; SE^{b}S. 90′; S^{h}W. 80′.
Variation = 8° E.
Required the latitude and longitude arrived at. In this example, an island is in lat. 39° 20′ S. and long. 21° 50′ W., and a second island bears from the first W. ½ N. (mag.) distant 20 miles. A *sunken* rock lies 10 miles to the westward of the second island with the two islands exactly in line. Place the islands and the sunken rock upon the chart, and state how near to the rock the ship passed.   (Oct. 1877, and again in March, 1879.)

(17) Mercator's chart, 1·85 in. = 1° long.
To extend from lat. 2° 30′ N. to 2° 30′ S. and long. 0° to 4° W.
A ship sailed from lat. 1° 45′ S., long. 0° 30′ W., as follows:
N. 49° W. 80′; N. 33° E. 85′. Variation = 17° W.
A second ship sailed from lat. 1° 50′ N., long. 3° 15′ W., as follows:
S. 13° E. 90′; S. 48° E. 95′. Variation = 17° W.
Protract the true courses, giving the latitude and longitude of the position at which the tracks cross, and the *true* bearing and distance of this position from the starting-point of each ship.
(Feb. 1878.)

(18) Plane chart. (Latitude and longitude equal)
1·75 inches = 1 degree.
To extend from lat. 1° to 6° S. and long. 19° to 24° W.

A steamer in lat. 1° 50' S., long. 20° 45' W., receives information that 4 days previously a dismasted ship had been seen in lat. 5° 30' S., long. 21° 20' W. Variation = 5° W., current = W^{b}N. 15 miles a day. Mark upon the chart the position of the dismasted ship at the time the steamer received this information, and state what magnetic course the steamer ought to steer to reach her; state also in what lat. and long. the steamer may expect to find her, if she start at once for the vicinity, and steam at the rate of 10 knots an hour.   (March, 1878.)

(19) Mercator's chart. Scale 1 in. = 5° longitude.
To extend from lat. 20° to 50° S. and from long. 10° to 60° W. Meridians and parallels to be ruled at intervals of 10 degrees.

Ship sailed from lat. 34° 53' S., long. 56° 10' W., direct to lat. 37° 8' S., long. 12° 10' W.

Where the track crosses the meridian 20° W., the variation = 15° W.

| ,, | ,, | 30° W. | ,, | = 11° W. |
| ,, | ,, | · 50° W. | ,, | = 9° E. |

Give the magnetic course at each point.   (April, 1878.)

(20) Mercator's chart. Scale 3 in. = a *mile* of longitude.
To extend from lat. 40° 10' to 40° 13' N. and long. 101° 10' to 101° 13' W.

A rock having less than 6 feet of water on it at low water average Spring Tides is situated in lat. 40° 11' 10" N. and long. 101° 11' 50" W. Variation = 11° W.

Place the rock on the chart, and give its magnetic bearings and distances from each of the inside corners of the margin of the chart.   (Oct. 1879.)

(21) Mercator's chart. Scale 1·1 inch = 1° longitude.
To extend from lat. 42° 20' S. to 48° 00' S. and long. 80° 00' E. to 86° 30' E.

Taking departure from a position with lighthouse in lat.

43° 00′ S. and longitude 82° 45′ E. bearing west (true) distant 25 miles, a vessel sails as follows:—

Comp. co.	Dev.	Dist.	
SW. ¼ W.	7° W.	118′	
EᵇS.	12° E.	155′	Variation = 21° E.
N. ¼ W.	3° W.	158′	

How does the lighthouse bear on the completion of the last course ? (Sept. 1880.)

(22) Mercator's chart. Scale 2·72 in. = 1° longitude.
To extend from lat. 61° to 62° N. and long. 18° to 22° E.
A ship sailed from lat 61° 10′ N., long. 18° 13′ E.
Soundings at the end of each distance :—

Mag. co.	Dist.	Fms.	
NNE.	40	30 m.	
SEᵇS.	52	47	Variation = 8° W.
NᵇE.	36	105	
SE. ¼ S.	48	70 m.	

Observations then placed the ship in lat. 61° 30′ N., long. 21° 12′ E. Allowing for current, and supposing the speed uniform, correctly place the soundings on the chart. (Nov. 1875.)

(23) Mercator's chart. Scale 3·04 in. = 1° longitude.
To extend from lat. 39° to 40° 40′ N., long. 124° to 127° W.
A ship sailed from lat. 39° 10′ N. and long. 126° 48′ W. as follows.
Soundings tried for at the end of each distance.

Mag. co.	Dist.	Fms.	
North	68′	$\overset{\cdot}{100}$	
SE. ½ E.	49	$\overset{\cdot}{120}$	Variation = 17° E.
NNE.	52	$\overset{\cdot}{100}$	
SEᵇS.	59	$\overset{\cdot}{120}$	

Observations placed the ship in lat. 39° 24′ N. and long. 124° 20′ W.

Allowing for the current experienced, and assuming a uniform speed, correctly place the soundings. (Aug. 1876.)

(24) Mercator's chart. Scale 3·6 in. = 1° longitude.
To extend from lat. 56° to 57° N. and long. 0° to 3° E.
A ship sailed from lat. 56° 15′ N. and long. 2° 40′ E. as follows.
Soundings at end of each distance:

Mag. co.	Dist.	Fms.	
North	30′	40	
SW. $\frac{1}{2}$ W.	35′	45	Variation = 20° W.
N. $\frac{1}{2}$ W.	44′	47	
SW.	40′	47 m.	

Observations then placed the ship in lat. 56° 12′ N. and long. 0° 24′ E.

Allowing for the current, and assuming a uniform speed, place the soundings in their correct positions. (Dec. 1876.)

(25) Mercator's chart. Scale 3·7 in. = 1° longitude.
To extend from lat. 57° N. to 58° N. and long. 1° to 4° E.
A ship sailed from lat. 57° 12′ N. and longitude 1° 25′ E. as follows.
Soundings at the end of each distance:

Mag. co.	Dist.	Fms.	
N^bE. $\frac{3}{4}$ E.	30′	60	
SSE. $\frac{1}{2}$ E.	47	65	Variation = 20° W.
NNE.	37	78	
SSE.	38	89	

Observations then placed the ship in lat. 57° 26′ N. and long. 3° 35′ E.

Assuming a uniform speed, and allowing for the current, place the soundings in their correct positions. (April, 1877.)

ANSWERS.

1°. Lat. 50° 20′ N., Long. 20° 46′ W.
2°. Lat. 61° 28′ N., Long. 177° 14′ W.
3°. Lat. 65° 3′ N., Long. 3° 42′ W.
4°. Lat. 49° 48′ N., Long. 8° 25′ W.

5°. Lat. 55° 46' N., Long. 19° 0' W. Current NE. (true) 37'.

6°. Lat. 57° 26' N., Long. 3° 44' E. Current NW^{b}W. (true) 11½'.
Mag. co. E. ¼ S.

7°. Lat. DR. 38°29' S., Long. DR. 79° 16' E.⎫ Current E. ½ N.
„ Obs. 38°27' S., „ Obs. 79° 27' E.⎭ (true) 9'.

8°. True co. NE.; Mag. co. ENE. 186'; Comp. co. E^{b}N. ¾ N.

9°. Lat. DR. 59°41' N., Long. DR. 4° 36' E. ⎫ Current S. ½ W.
„ Obs. 59°26' N., „ Obs. 4°33' E. ⎭ (true) 16'.

10°. Lat. DR. 51° 6' N., Long. DR. 10°22' W.⎫ Current NW^{b}W.
„ Obs. 51°10' N., „ Obs. 10°32' W.⎭ (true) 8'.

11°. Lat. DR. 51° 9' N., Long. DR. 9°44' W.⎫ Current NE. ¼ E.
„ Obs. 51°20' N., „ Obs. 9°17' W.⎭ (true) 21'.

12°. Lat. DR. 48°25' N., Long. DR. 5°28' W.⎫ Current W^{b}N. ¾ N.
„ Obs. 48°26' N., „ Obs. 5°35' W.⎭ (true) 5'.

13°. Lat. DR. 64°49' N., Long. DR. 19°15' W.⎫ Within 8½ miles of
„ Obs. 64°48' N., „ Obs. 19°18' W.⎭ Islet.

14°. Lat. DR. 50° 3' N., Long. DR. 4° 3' W.⎫ Comp. co. E. ¾ S.
„ Obs. 50° 4' N., „ Obs. 4° 2' W.⎭

15°. Lat. DR. 36°37' S., Long. DR. 72°12' E.⎫ Current NE. 4'.
„ Obs. 36°35' S., „ Obs. 72°14' E.⎭

16°. Lat. DR. 39°38' S., Long. 23° 21' W.
2nd Id. is in Lat. 39° 16' S., Long. 22° 16' W. Rock in 39° 14' S., 22° 28' W. Ship passed within 30' of the Rock.

17°. Lat. of Point 0° 12' N., Long. 1° 39' W.
From 1st Ship's point of departure NNW. ¾ W. 124'; from 2nd Ship's do. SE^{b}S. ¾ E. 125'.

18°. When intelligence is received, wreck is in 5° 24' S., 22°20' W. It will be found in 5° 22' S., 22°37' W. Mag. co. SSW. ¾ W.

19°. In Long. 50° W. Mag. co. E. ½ N.; in 30° W. E^{b}S. ¼ S.; in 20° W., E^{b}S. ½ S.

20°. From NE. corner S. 60° W. 2'·4; NW. corner S. 26 E. 1'·9; SW. corner N. 35 E. 2·7; SE. corner N. 26 W. 3'·02.

7—2

MARINE SURVEYING.

21°. Lt. Ho. bears W. ½ N. (true) 79'. Lat. in 43° 10' S., Long. in 84° 33' E.

22°. Lat. DR. 61° 15' N., Long. 21° 7' E.; N♭E. 15'.

   1st Sounding in Lat. 61° 53' N.; 2nd in Lat. 61° 19' N.; 3rd in 61° 57' N.; 4th in 61° 30' N.

23°. Lat. DR. 39° 18' N., Long. 124° 42' W.

   1st S. in 40° 17' N.; 2nd in 39° 38' N.; 3rd in 40° 18' N.; 4th in 39° 24' N.

24°. Lat. DR. 56° 17' N., Long. 0° 39' E.

   1st S. in 56° 24' N.; 2nd in 56° 11' N.; 3rd in 56° 49' N.; 4th in 56° 12' N.

25°. Lat. DR. 57° 19' N., Long. DR. 3° 22' E.

   1st S. in 57° 44' N.; 2nd in 57° 11' N.; 3rd in 57° 51' N.; 4th in 57° 26' N.

Note. The above Results have been found by actually drawing the Charts, and it is hoped that they are well within the limits of error which may naturally be looked for in work of the kind.

CHAPTER VI.

INSTRUMENTS AND OBSERVING.

76. GUNTER'S CHAIN.

This is divided into 100 links, each link being 7·92 inches long; hence the whole chain is 66 feet, or 22 yards, or 4 perches.

$\therefore$ 1 square chain = 16 perches = $\frac{1}{10}$ acre.

The great advantage of this chain is the facility with which areas may be computed from the measured lengths of the sides.

77. THE ORDINARY SURVEYING CHAIN is 100 feet in length. It is composed of steel bars and chain links, each bar and the three chain links being equal to one foot. The chain is divided into segments of 10 feet from the middle towards each handle. These segments are marked by small pieces of brass so divided at their edges that the number of 10 feet from each handle can be detected at a glance. Swivel joints at the handles and at the middle prevent the chain getting twisted.

In measuring with a chain every precaution must be taken, as the results are liable to many errors arising from (1) the chain itself, and (2) the method of using it.

If the chain is stretched too tight the links give, and therefore the measured length is shorter than the real length; if not stretched tight enough the measured length will be too short. If the chain is a new one a comparison ought to be made at the close of the day's work; if however it has been in use for some time and found trustworthy, a comparison once every three or four days will suffice.

The method of using the chain in measuring a line is fully described in the chapter on Base Lines.

78. Sextant.

There is no necessity to occupy space in the actual description of this instrument. Hadley invented this invaluable aid to navigation about the year 1731; the principle of the sextant however appears to have been known to Sir Isaac Newton. We shall merely notice the ADJUSTMENTS in their proper order.

79. I. The index glass must be perpendicular to the plane of the instrument.

II. The horizon glass must be perpendicular to the plane of the instrument.

III. When the horizon glass is parallel to the index glass the index error ought to be zero.

IV. The line of collimation must be parallel to the plane of the instrument.

80. To test these Adjustments.

I. *Is the index glass perpendicular to the plane of the instrument?*

Place the moveable radius called the index bar about the middle of the arc, or somewhat nearer to the commencement of the graduated limb. Then look obliquely into the index glass. If the limb and its reflected image appear in the same line the adjustment is correct; if not, it must be looked to. If the image is *too high*, the glass leans forward; if *too low*, the glass leans backward.

Note. This correction is carefully made before the instrument leaves the maker's hands, and is not easily put out of adjustment.

II. *Is the horizon glass perpendicular?* No. I. adjustment is first made, because the index glass being perpendicular, if in

VI.] INSTRUMENTS AND OBSERVING. 103

any position it is parallel to the horizon glass, the latter must also be perpendicular to the plane of the instrument. We test the accuracy of No. II. in two ways.

(α) By making the reflected image of the sun pass over the sun looked at through the telescope. If the two bodies pass on one side of each other the adjustment must be seen to.

(β) By the sea horizon. Set the index about zero, and look directly at the horizon through the unsilvered part of the glass. If the reflected image is in line with this horizon and remains so while the sextant is turned through a very large angle, the adjustment is correct; if not, it must be seen to.

A little practice will enable most persons to remedy this source of error by means of the screws behind the horizon glass.

III. *Is the index-error zero?* It is not *necessary* that it should be so, but it is necessary that its amount should be known, and allowed for in all computations. When the adjustments are made in the above order, the position of the index of the vernier with respect to the zero of the limb may be determined in three different ways. (α) By making the reflected image of a star or sun to coincide exactly with the object looked at directly through the telescope; (β) by measuring the sun's diameter on and off the arc; (γ) by the sea horizon.

IV. *Is the line of collimation correct?* Make a contact at one of the wires of two stars about 100° or 110° apart. Throw the stars on the other wire: if the contact is still maintained the adjustment is correct; if not, it may be adjusted by the following practical rule[*].

" Open the screws of the collar of the telescope on that side on which the separation of the stars takes place."

Note. If the separation takes place on the wire *farthest* from the plane of the instrument the object end of the telescope *droops towards* the instrument, and if the separation takes place at the nearer wire, then the eye-piece droops.

[*] Communicated by Staff Commander V. Johnson, R.N.

If this adjustment is not correctly made, then the middle of the wires is no longer the true line of sight, and the contact observed there will give angles which are too great*.

To obtain a good observation with the sextant the following points ought to be regarded: (1) the images ought to be sharply defined by careful focussing, (2) when observing the sun, *neutral* tints ought be used as less fatiguing to the eye, (3) in using the artificial horizon the suns ought to be of the same brightness.

Note. In observing the angle between the moon and a star, make the edge of the moon to bisect the star's light; and in the case of the moon and planet, bring the edge of the moon's disc to the estimated centre of the planet.

81. THEODOLITE.

The following description of this instrument will be clearly understood if the student keeps the annexed diagram of a Theodolite before him for reference.

The three principal parts are, the Levelling Plates, the Horizontal or Azimuth Limb, and the Vertical Arc. We shall notice these in detail.

I. THE LEVELLING PLATES. These (marked A and B) are held together by a ball and socket joint, and can be set firm and parallel to each other by means of four milled-headed screws (S): these screws turn in sockets fixed to the lower plate (A) while their heads press against the under side of the upper plate (B). In some instruments this order is reversed. The screws are set in pairs, and when made to act in contrary directions the instrument is set up level.

II. THE HORIZONTAL OR AZIMUTH LIMB consists of two circular plates, the upper (V) called the *vernier circle* and the lower (C) known as the *graduated circle*. The latter projects somewhat beyond the former, the two surfaces being conical and nicely fitted

* Vide Jean's *Navigation*, Part I. page 113, for a method of computing the correct angle from the observed angle. This source of error is more frequent than is usually suspected.

to each other, the upper having a perfectly easy and very steady motion along the lower. Both circles have horizontal motion round the central axis. This axis consists of two parts, an outer and an inner. The vernier circle is attached to the *inner*, and the graduated circle to the *outer*. At opposite points of the vernier plate a short space is chamfered, forming with the graduated circle an even surface. These small spaces contain the verniers. The graduated circle is divided into 20′ spaces*, and 39 of these are taken for the length of the vernier, which is divided into 40 equal parts: therefore on such an instrument we can read to *half minutes*. In smaller instruments we can read only to single minutes.

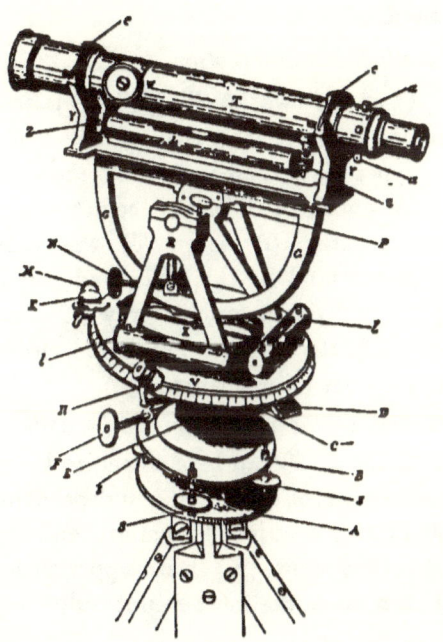

III. THE VERTICAL ARC. Attached to the vernier circle are two frames (R) which support the pivots of the vertical arc or semicircle (G), on the top of which is placed the telescope (T).

* This is only one example. The arcs seem to be graduated in many different ways.

On one side an arm is attached to the axis of the vertical arc, and carries at its other end a microscope for reading off the angles of elevation or of depression. The vernier is fixed, and this microscope can be moved along its face for the purpose of reading off.

82. We next come to speak of the several MOTIONS of the Theodolite. These are three in number, viz.—

I. *The absolute horizontal motion of the whole instrument about its axis.*

II. *The motion of the vernier circle with respect to the graduated circle.*

III. *The motion of the vertical arc.*

Details of these motions:

I. *The absolute horizontal motion.*

The lower (A) of the two levelling plates is screwed down to the legs of the instrument; the axis of the instrument passes through to the upper plate, and is fixed at the other end to the vernier circle. When the two circles composing the horizontal limb are clamped together (by screw K), the whole instrument can be made to turn round on this axis, and can be clamped in any required position by a clamping screw (D) situated between the upper levelling plate and the graduated circle. A tangent screw (F) gives the finer adjustments.

II. *The motion of the vernier circle in azimuth.*

The graduated circle (C) can be clamped to the axis as above explained. The vernier circle, having independent motion, can then be turned in any required direction, and can be fixed at pleasure by a clamping screw (K) in its upper surface: a tangent screw (M) will then make the more minute adjustments.

III. *The motion of the vertical arc.*

The vertical arc (G) can be clamped in any position by means of a screw (P) which acts on the horizontal axis; a tangent screw (N) at the lower part of the arc moves the arc and the attached telescope through very minute intervals, and thus the observation can be made accurately.

83. Before speaking of the adjustments and the methods of making them, we may observe that attached to the vernier circle (V) are two small spirit-levels (l, l), placed at right angles to each other. These are intended to secure the horizontal position of the azimuth limb. Another level (L) is situated immediately below the telescope; the intention being to ensure the axis of the telescope being horizontal if required. A compass (X) is fixed on the vernier circle to enable approximate bearings to be taken. A plummet attached to a hook immediately beneath the axis enables us to place the vertical axis of the instrument exactly over a given point.

Coloured eye-pieces are also supplied for the observations of the sun.

84. The ADJUSTMENTS OF THE THEODOLITE are three in number.

I. *The line of collimation must be correct.*

II. *The spirit-level beneath the telescope must be parallel to the line of collimation.*

III. *The axis of the instrument must be truly vertical.*

First adjustment. The line of collimation must be correct: i.e. the line joining the centre of the object-glass with the intersection of the cross wires must coincide with the axis of the rings (Y, Y) on which the telescope rests, i.e. with the "*line of the Y's.*" To ascertain whether this is so or not, we must look through the telescope and cause the intersection of the wires to "bisect" some distant and well-defined point; then turn the telescope upside down, so that the spirit-level is now at the top of the telescope, and observe whether the wires still bisect the object. If so, the adjustment is correct; if not, we must move the circular frame carrying the wires* through half the deviation by turning two of the small screws (releasing one before tightening the other) which keep the diaphragm or frame in its position, and the other half must be corrected by elevating or depressing the telescope. If necessary this operation must be repeated until the

---

* See diagram on p. 111.

adjustment is perfect. A similar operation will correct the other wire if required. The small screws (*a, a*) of the diaphragm appear on the outside of the telescope.

Second adjustment. The spirit-level attached to the telescope must be parallel to the line of collimation. The clips (*c, c*) which retain the telescope in its Y's (as the cylindrical rings on which it rests are called) being opened, and the vertical arc clamped, bring the bubble to the centre of its "run" by means of the tangent screw (*N*) of the vertical arc. This done, carefully turn the telescope end for end in its Y's, so as not to disturb the vertical arc; if the bubble resumes its position in the centre of its run, the adjustment is correct; but if not, then it must be brought back one half of the distance it has moved by means of *the screw at one end of the level* (*Z*), and the other half by the *tangent screw of the arc* (*N*). This operation must be repeated until the adjustment is perfect.

Third adjustment. The axis of the instrument must be truly vertical, or what amounts to the same thing, *the horizontal limb must be truly horizontal.*

Set the instrument as nearly level as possible by the eye. Clamp the graduated circle (by *D*), but let the vernier circle be free. Move the latter until the telescope is over two of the levelling screws (*S, S*); then bring the bubble of the telescope level (which is the most sensitive in the instrument) to the middle of its "run" by means of the tangent screw (*N*) of the vertical arc; next turn the vernier circle through 180 degrees, when if the bubble returns to the middle, the limb is horizontal in that direction; but, if not, half the difference must be corrected by *the levelling screws over which the telescope is lying,* and half by means of *the tangent screw of the vertical arc*. Having done this, turn the vernier circle through 90 degrees, so that the telescope may lie over the other pair of parallel screws; and by their motion make it horizontal. Having thus made the azimuth limb horizontal by means of the sensitive telescope-level, the bubbles of the other levels on the vernier circle must be brought to the centres of their "runs" by the screws which fasten them in their places.

The vernier of the vertical arc may now be attended to. The index error (i.e. the deviation of the arrow head from zero, when the Theodolite is perfectly adjusted) is best obtained as follows:— repeat the observation of an altitude or depression in the reversed positions both of the telescope and the vernier circle. The two readings will have equal and opposite errors; the index error is half their difference.

85. Method of using the Theodolite.

Adjust the screws (S, S) between the levelling plates so that equal lengths may appear above the upper plate. Extend the legs of the instrument until the bubbles of the levels on the vernier circle are nearly in the middle of their "runs," and the plummet hangs freely over the required point, and press the legs firmly in the ground. Then unclamp (D) the whole instrument so that it may turn freely about its axis, but keep the other motions clamped (K, P). Adjust the instrument so that the azimuth limb may be horizontal. Clamp the whole instrument (D), and unclamping (K) the two circles of the azimuth limb, set the index of the vernier to 360° or *zero* of the graduated circle and clamp it; examine the other vernier, the index of which ought to coincide with 180°. This must be done by means of the attached microscope.

Next loosen the large clamping screw (D) and turn the whole instrument towards the *left* of the two stations between which the first angle is to be taken; bisect this object as closely as possible by hand: then firmly clamp the instrument and make the observation exactly by aid of the tangent screw (F). Now as the index points to zero, and the lower circle is graduated from left to right, it is evident that, by separating the vernier circle and turning it round to the right until the second object is bisected, the angle can be read off between the first and second object. Both verniers must be read and the *mean* taken.

In observing angles with a Theodolite the following method is adopted in the Ordnance Survey.

Let A, B, C...K, be the points observed taken in order of azimuth; then the instrument being in perfect adjustment A is bisected and the microscope read, then *B*, then *C*, and the other points in succession; after observing *K*, the movement of the telescope is continued *in the same direction* round to *A*, which is observed a second time, to ascertain whether the instrument has moved. This complete round is termed " an arc." A more ordinary procedure is to observe the points as before in the order *A*, *B*, *C*, ...*K*, then reversing the direction of motion of the telescope to re-observe in the inverted order, *K*, *H*, ...*C*, *B*, *A*. Thus each point is observed twice.

86. TO REPEAT AN ANGLE. After the last observation, without detaching the two circles turn the whole instrument round to the first object, and then unclamping the vernier circle turn it round until the second object is "bisected." The difference between this and the first reading will be *double the mean angle*. Again, keep the two circles together, turn round as before to the first station, and then the vernier circle on to the second object; the difference between this reading and the first reading will be *three times the mean angle*. &c.*

87. TO OBSERVE A VERTICAL ANGLE. Level and adjust the instrument. Bring the bubble of the telescope-level to the centre of its " run." Make the zero of the arc coincide carefully with the index of the vernier. Then " bisect " the object, and the changed position of the broad arrow will mark the angle of elevation or depression. Reverse in azimuth, also reverse the telescope, and take the mean of the two readings if they are different†.

88. THE SPIRIT-LEVEL.

This is also called the " Y-level" from the supports on which the telescope rests being shaped somewhat like the letter Y. The most perfect instrument of this kind is called the Improved Dumpy Y-level; the term " dumpy " being given because of its short length and large telescope.

---

* The first object observed in the arc, or round of angles, is termed the *zero*, because where "bisected," the index points to zero on the graduated circle.

† Or we may apply the Index Error to the observed altitude or depression, *vide* p. 109.

Description of the instrument.

The Spirit-Level consists of a telescope resting on two supports shaped like the letter Y, these supports being known as the Y's. The instrument may be set level exactly in the same way as the azimuth limb of the Theodolite (vide the *third adjustment* of that instrument). One of the Y's can be moved in the same way that the telescope of the sextant is moved in its socket. The instrument is clamped by means of a collar (E in diagram of the Theodolite) which surrounds the vertical axis. Beneath the telescope is fixed a sensitive level, attached by a hinge at one end to the telescope, while a capstan-headed screw at the other end tends to raise or lower that end. Two spider lines at right angles give the centre of view. When this point of intersection is *on* with any object, the object is technically said to be "bisected."

ef is a brass ring somewhat smaller than the tube of the telescope. Two fine lines of spider's web are fixed to this ring at right angles to each other. By loosening a screw at D and tightening the screw at C, the ring can be moved from right to left, and similarly up or down as required. These screws project beyond the side of the telescope and can be moved without difficulty.

Before adjusting the focus of the object-glass that of the eye-piece ought to be looked after, both in this instrument and in the

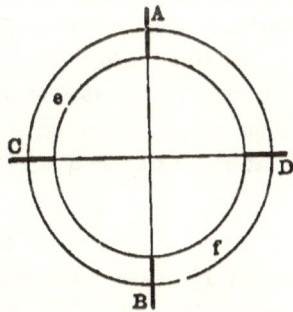

Theodolite. The eye-piece must be drawn out until the cross wires are clearly defined, and there is no instrumental parallax, i.e. on looking at some distant object and bringing the intersection

of the wires *on* with it, there may be no displacement of the contact on moving the eye to the left or right. Such a parallax would be caused if the image of the object falls beyond or falls short of the cross wires. Its presence can always be detected by moving the eye about and noticing if the cross wires change their position, or are fluttering and indistinct. Instrumental parallax must be corrected as follows:—

1st. Adjust the eye-piece until the cross wires are sharply defined against any object.

2nd. Thrust forward the object-glass by means of the screw at the side of the telescope to get a correct focus (W in diagram).

3rd. Then readjust the eye-piece if necessary.

The parallax must be corrected before we look to the collimation.

The three following ADJUSTMENTS are made in the Spirit-Level.

I. *The line of collimation must be parallel to the axis of the Y's on which the telescope is lying.* It is corrected as in the case of the Theodolite; vide the *first adjustment* of the Theodolite.

II. *The spirit-level must be parallel to the line of collimation.* After the bubble has been brought into the centre of its "run" by the plate screws, the telescope is reversed in its supports (i.e. turned end for end); if the bubble has moved, it must be brought back to the centre through one half of its displacement *by the screw at the end of the level*, and through the other half *by the plate screws*. This will require several repetitions, and then the clips are secured in their places. The object of the adjustment is to make certain that the axis of the telescope is truly horizontal.

III. *The Y-supports must be exactly on the same level*, so that when adjustments I. and II. have been made the axis of the telescope may revolve in a plane at right angles to the axis of the instrument. This may be effected as follows:—

Level the telescope when placed over two of the levelling screws, and then reverse the telescope; if the bubble has moved it must be adjusted by bringing it back through one half of its displacement by *turning the capstan-headed screw* placed directly below one of the Y's, which can thereby be raised or lowered in its socket, and through the other half *by the plate screws*. This operation must then be repeated with the other pair of plate screws. After a few trials it will be found that the bubble remains in the centre of its "run" as the telescope revolves completely round the axis of the instrument.

89. THE LEVELLING STAFF.

This is usually made in three pieces which slide into one another in the manner of a telescope. When drawn out a spring at the lower end of each holds it in position, and thus a very convenient staff is formed 14 or 17 feet long. The lowest joint is about 4 inches wide, the next about $3\frac{1}{4}$ inches, and the top about $2\frac{1}{2}$ inches. The entire length is divided into feet, and these are again subdivided into tenths and hundredths. These smallest divisions are coloured black and white alternately. The numbers representing the *feet* are sometimes painted *red*. These numbers are placed on the left of the graduations, and the *tenths* on the right. As the divisions are very small every advantage is taken to distinguish the marks to facilitate the reading off through a telescope at some distance; e.g. the top and bottom edges of the horizontal stroke of the figure 7 will coincide with divisions, and the lower edge of the long stroke of the same figure will be in a line with a division; while the whole figure will occupy ten of the hundredth divisions, or one-tenth of a foot. All these minutiæ are of assistance when reading off at a distance of 300 or 400 feet. When looked at through the telescope of the level the staff appears inverted. It may be noted that when the assistant holds the staff he sways it gently to and fro in the direction of the level: the *least reading* will of course be when the staff is held upright. The vertical wire in the telescope will be the criterion that there is no lateral deviation in the position of the staff.

90. THE TEN-FEET POLE.

This consists of two vanes about 2 feet by 15 inches, which slide on the ends of a pole about 13 feet long. On the vanes are painted two lines at right angles to the pole. These vanes are fixed to the pole so that the distance between the centres of the lines is exactly 10 feet. This instrument is used for putting in the coast line. The following explanation will shew the method of using it.

The scale *for the plan* must be first determined*. A side of any triangle is computed from the measured base line, and the length of this side is taken from the plan. Then the length of the nautical mile for the given latitude being known, we get the following proportion:—

$$\frac{\text{Scale for 1 mile}}{\text{length of the nautical mile}} = \frac{\text{measured distance from the plan}}{\text{computed length of side}}.$$

e.g. Let the distance between two points on a plan be 7·5 inches, and let the computed distance be 9120 feet; then, if the mile be 6080 feet, we shall have

$$\frac{\text{Scale for 1 mile}}{6080} = \frac{7 \cdot 5}{9120}.$$

∴ Scale for 1 mile = 5 inches.

This plan would therefore have been projected on a scale of 5 inches to a mile, or 1 inch to 1216 feet.

∴ A distance in inches on the plan = $\dfrac{\text{Computed length in feet}}{1216}$.

The Scale for the Ten-feet Pole can now be constructed as follows:—

Let AB represent 10 feet, viz. the distance between the centres of the vanes, and let C be the point of observation; suppose the angle ACB subtended by the length AB to be θ (*Note*. In practice this angle is *always* measured *on and off the arc*, and the mean angle taken):

Then $\tan\dfrac{\theta}{2} = \dfrac{AD}{CD};$

---

* *Vide Notes on Marine Surveying*, by Staff Commander V. F. Johnson, R.N.

INSTRUMENTS AND OBSERVING.

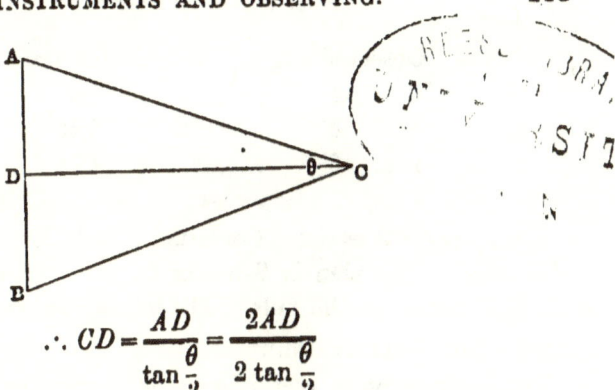

$$\therefore CD = \frac{AD}{\tan\frac{\theta}{2}} = \frac{2AD}{2\tan\frac{\theta}{2}}$$

$$= \frac{2AD}{\tan\theta} \text{ nearly};$$

$$\therefore \text{computed distance} = \frac{10 \text{ feet}}{\tan\theta};$$

$$\therefore \text{distance in inches on the plan} = \frac{\text{Computed distance}}{\text{scale for 1 inch on the plan}}.$$

Suppose the scale to be 5 inches to a mile of 6080 feet

$$\therefore \text{distance on plan} = \frac{10}{1216\tan\theta}.$$

Let
$\theta = 1'$,
log 10 = 1·000000
log cot 1' = 13·536274
           14·536274
log 1216 = 3·084934
$\therefore$ log distance = 1·451340
$\therefore$ distance = 28·27 inches.

Hence if the observed angle be *one minute*, the distance of the pole from the point of observation will be 28·27 inches on the plan. Now we can assume, without sensible error, that when the angles are very small, at half the distance the angle will be twice as great, or conversely; hence the *scale for the* 10-*feet pole* may be constructed as follows:—

Observed angle.	Inches of scale.
1'	28·27
2'	14·13
3'	9·42

Observed angle.	Inches of scale.
4'	7·06
5'	5·65
6'	4·71
&c.	

Examples for Exercise. Construct a scale for the 10-feet pole if the scale of the plan is 8 inches to the mile, and also if the scale is 6 inches to the mile. The mile = 6080 feet.

91. THE STATION POINTER.

This instrument is useful in fixing a point by means of the "Three-point problem." We may give the following description of its construction and use.

The instrument consists of a ring 6 inches in diameter, and about ⅝ inch wide*. From a central ring proceed three arms, about 12 inches long, and these can be increased to 18 inches by extra pieces supplied for the purpose. The central arm is fixed, its bevelled edge coinciding with the zero of the graduations. This graduation is continued to the right and left from the zero. The other two arms can be moved through any required angle, and can be fixed by a clamping screw. A tangent screw on the side will give the finer adjustments. Each arm carries a vernier, and by its aid we can lay off angles to single minutes.

One edge only of each arm is bevelled; if the right-hand edge is bevelled, the instrument is a "Right-hand Station Pointer," and similarly for the left hand. One of these arms can close up nearer to the index radius than the other. Hence if the left-hand angle is smaller than the left radius can measure we must have resort to a right-hand pointer, and *vice versâ*.

A small stroke at the centre of the inner ring marks the point of observation required.

92. METHOD OF USING THE INSTRUMENT.

The angles observed are set by means of the verniers, and the bevelled edge of the left-hand radius bar is placed over the left-hand object, and then placing the thumb and fore-finger on either side of the bar, the instrument is moved about until the bevelled edge of the middle

---

* These instruments are made of all sizes.

bar concides with the middle object, when it will be found easy to make the edge of the third bar coincident with the third object. A pricking point will then mark the position required.

93. THE BAROMETER.

This instrument is intended to shew the pressure of the air at any moment. In its simplest form it merely consists of a glass tube, somewhat less than 3 feet long, closed at one end, which is filled with pure mercury, and with the other end inserted in a vessel containing mercury. When held vertically, the column of mercury inside the tube will exactly balance the pressure of the air on a section equal to that of the tube. In general terms, when the level of the mercury *rises*, the air is exerting an increased pressure, when it *falls*, the air is exerting a diminished pressure. It does not fall within the scope of this work to give the methods of *interpreting* the records of the barometer. These are fully explained in various publications, and more particularly in the *Barometer Manual* published by the Board of Trade*. From that excellent little work most of the following details are taken.

Barometers are best mounted in brass, because its coefficient of expansion by heat is well known, and the tables for correcting barometer readings for temperature (founded upon the coefficients of expansion of mercury, glass, and brass) always give with such barometers identical results. Tables have been formed for barometers framed in different kinds of wood, but, for accurate results, these instruments cannot be relied on.

If, however, and this is important, a scale be applied which is quite independent of the frame, then the reduction for temperature will depend upon the material of the scale, and in such a case wood will answer perfectly as a frame. The scale may be of ivory, porcelain, or enamel, and is fixed in its proper position.

It is evident that if the barometer rises, the increased quantity of mercury in the tube must have come from the cistern, and therefore the surface of the mercury in the cistern must be lower in consequence; and the opposite result follows if the barometer

* J. D. Potter, 31, Poultry. Price One Shilling.

is falling. Hence the varying level of the surface of the cistern must be corrected or allowed for. There are *three* methods of doing this: (1) by *capacity correction*, (2) by a *flexible base to the cistern*, and (3) by a *contracted scale*.

94. (I.) *By capacity correction.*

When the cistern is covered up and the scale is engraved on the frame this method is adopted :—A certain height of the column in the tube is correct by the scale, and the position of the mercury at this time is known as the *Neutral Point*. When the mercury sinks below this point, the level rises in the cistern above the zero of the scale, and hence the reading will be *too great*. When the mercury rises above this point, the level in the cistern falls below the zero, and hence the reading is *too small*. On such a barometer the neutral point is marked (N.P.), and also the relative interior sections of the tube and cistern thus, "Capacity 1 to 50." From these elements, the Correction for Capacity is found by taking $\frac{1}{50}$th of the difference between the height read off and that of the N.P., *adding* the correction to the reading when the column is above, and *subtracting* it from the reading when it is below the N.P.

95. (II.) *By a flexible base to the cistern.*

This is the principle of Fortin's Barometer. The upper part of the cistern is made of glass, the base is flexible and is acted upon by a lifting screw placed beneath. From the top of the cistern an ivory cone descends, the point of which is the zero of the scale. Before reading off, the level of the mercury in the cistern is brought to this point by means of the lifting screw, and then the height of the column in the tube as read off from the scale will be the true reading.

This construction is usually adopted in standard barometers, and cannot be used at sea.

96. (III.) *By contracted scale.*

In this the highest point of the scale is the neutral point, and the inches of the scale are shortened in proportion to the relative size of the sections of the tube and of the cistern. Thus, if the diameter of the tube be 0·25 in. and that of the cistern be 1·25 in. then the inches on the scale are shortened by 0·04 of an inch.

97. THE MARINE BAROMETER.

The greater part of the length of the tube of a marine barometer must be made with a very fine bore, to prevent oscillations in the column of mercury from the rolling of the ship. When the bore is not sufficiently contracted, the fluctuations which arise from the motion of the ship, are called "pumping." If on the other hand the bore is too contracted, the instrument is sluggish in responding to the varying pressure of the atmosphere, and is therefore not suitable for very accurate observations.

Whenever a marine barometer ceases to act, and there are no traces of damage to the instrument, it may be surmised that a particle of dirt, or a bubble of air, has lodged in the very fine contraction of the bore. To remedy this defect the instrument should be taken down, the mercury allowed to fill the tube, and put aside in an *inverted* position for a few hours. When replaced in its proper position the cause of error will generally be found to have disappeared.

The marine barometer is mounted in a brass frame, but the cistern is of iron. The frame is open in front and rear to expose the "range portion" of the tube; the scale is protected from dust by a glass shield. The vernier is engraved on a piece of silvered brass tubing, and travels firmly by a rack and pinion motion, the parts being kept in position by friction. The inches of the scale are contracted as explained above.

The cistern is closed. It contains sufficient mercury to cover the open end of the tube when the instrument is laid flat, or when inverted. A small aperture at the top, covered internally with leather, permits the pressure of the external air being exerted, but prevents the mercury escaping.

98. METHOD OF SUSPENDING.

The barometer must be hung in gimbals, and the arms of hammered brass by which it is supported ought not to be shorter than 12 inches to allow sufficient spring. The instrument must hang quite vertically by its own weight only, and especial care must be taken that the reading is not taken when in any other position, as the reading would be erroneous and too great. It ought to be kept in the shade, in a place easily accessible, and where it cannot come into collision with any object during the heaviest motion of the vessel.

99. Vernier of a Standard Barometer.

The scale is divided into inches, and each inch is subdivided into ten equal parts, these again being bisected, thus the inch contains twenty equal parts. Twenty-four of these are taken as the length of the vernier, and this space is divided into twenty-five equal parts, thus the Least Reading is

$$\frac{1}{25} \text{ of } \frac{1}{20} = \frac{1}{500} = \cdot 002 \text{ inch.}$$

In the case of the ordinary marine barometer used on board ship, the Least Reading is ·01 inch.

100. Method of Reading.

The lower edge of the vernier must be brought on a level with the slightly convex surface of the mercury, then this edge, the surface of the mercury and the back edge of the vernier (if there be a back edge) will be in one plane. White paper held behind will facilitate the accuracy of the reading.

101. Method of stowing for carriage.

The vernier must be brought down to the bottom of the scale: then the instrument must be lifted out of the sustaining bracket, and held for a few minutes in an inclined position, so that the mercury may flow *gently* up and fill the top of the glass tube. It must then be placed lengthwise in its box, with soft packing around it, and the lid *screwed*, not nailed, on. During the carriage it must be kept free from all jarring blows.

Note. The student is recommended to read carefully Raper's remarks on the precautions to be observed in taking observations. *Vide* his *Practice of Navigation*, Chapter III. pp. 175 to 184. Read also Chauvenet's remarks on the errors to which observations are liable. *Spherical and Practical Astronomy*, Vol. II. p. 471.

CHAPTER VII.

BASE LINES.

102. IF in any triangle we know the values of the three *angles* only, we can determine merely the ratios of the three sides, but if we know in addition the absolute value of any one side, then the absolute values of the other two sides may be determined. Hence the importance of determining in every survey the absolute length of one line accurately. The line so determined is known as the *Base Line* or simply the *Base*.

103. Now if in a triangle we know the length of one side and the two adjacent angles, the triangle can be completely solved. We can then use one of these calculated sides as the known side of a new triangle, and thus proceed indefinitely. Hence the necessity of measuring accurately the base line in the first instance, because if erroneous, the error goes on increasing with every triangle until finally the calculated sides become quite erroneous.

104. DEF. When a triangulation has been carried on over a country it is customary to measure one of the computed sides of a triangle, far distant from the starting point, with the same detail as the original base was measured, as a test of the accuracy of the survey. Such a test is the most severe to which the work can be put. This measured side is known as the "BASE OF VERIFICATION."

E. g. The Ordnance Survey of Great Britain and Ireland was carried on from a base measured on Salisbury Plain. When the network of triangles had reached the north-west of Ireland a base of verification was measured along Lough Foyle. The difference between the computed and measured lengths was less than 5 inches!

In 1793 a base of verification was measured by Mechain, near Perpignan, to test the accuracy of the work connected with the famous measurement of the Arc of the Meridian. The measured and calculated lengths differed by less than a foot, although the original base was distant more than 430 miles. Another instance of accurate work may be cited. When General Roy died in 1790, the English Ordnance Survey was carried on by General Mudge, from the measured base on Hounslow Heath, through Greenwich, to Dunnose in the Isle of Wight, thence on through Devonshire, Dorset and Wiltshire, and connected with the base of verification on Salisbury Plain. The computed and measured lengths were found to differ by scarcely 1 inch!

105. The measurement of a base line, although seemingly not difficult, is by far the most tedious and important part of a Trigonometrical survey; and hence in a very important survey, such as that of an entire country, every refinement which mechanical ingenuity can suggest has been lavished upon this operation to secure, as far as possible, mathematical accuracy.

We will first dwell briefly on the more *exact* methods of measuring a base used in such a survey, as likely to prove interesting to younger readers, and shall then enter more fully on the various methods used by nautical surveyors.

The geodetic standards of measurement in different countries vary in length, in form, and in the substance of which they are composed.

They may however be divided into two classes, known respectively as standards "*à traits*," and standards "*à bouts*." In the *former* class, the lines or dots defining the exact measure used are engraved on small disks of silver, platinum, or gold let into the bar; in the *latter*, the ends of the bar are generally in the form of a small cylinder presenting a circular disk, either plane or convex, of some hard polished metal, or agate, for the purpose of contact in the operation of measuring a distance.

Thus the standard used in the Ordnance Survey of Great Britain is 10 feet in length, and in section is a rectangle of $1\frac{1}{2}$ inches broad by $2\frac{1}{2}$ inches deep. It is supported on rollers at $\frac{1}{4}$ and $\frac{3}{4}$ of its length. The ends of the bar are cut away to half its depth, so that the dots marking the measure of 10 feet are in the neutral axis of the bar.

106. Standards are generally furnished with thermometers, which either lie in contact with the metal, or else have their bulbs so arranged as to fit into cavities in the upper surface of the bar. The errors of these thermometers must be known with great exactness, because an error of $\frac{1}{10}$ of a degree in temperature corresponds to an error of nearly a millionth of its entire length in an iron bar. These thermometers therefore are from time to time compared with the standard thermometers. Such a standard must be " the best workmanship of the best workman," and the residual errors (in the boiling point, freezing point, calebration, and comparisons of the standards) can be determined only by observation and experiment. Every thermometer has an index error which varies slightly in the course of time, and must be determined occasionally. When examined in the Ordnance Survey Establishment at Southampton, these thermometers are held in water on a small platform of perforated zinc with about 7 inches of water over them. There is a mechanism for slightly agitating the water and thus to prevent any local cooling. The instruments when lying horizontally are read from above by means of powerful micrometer microscopes.

107. The knowledge of the exact length of a bar at any moment involves three distinct matters :—

(1) *Its length at some specified temperature*, which is known by repeated comparisons with the standard bar.

(2) *Its coefficient of expansion*, which must be obtained from special experiments.

(3) *The temperature of the bar at the moment of observation*, which can only be discovered by means of the attached thermometers.

108. To evade the temperature difficulty, three forms of construction have been adopted.

(1) *Borda's.* In this, the measuring bar consists of two rods of metal having quite different rates of expansion: the slight expansions from the normal length being marked by verniers and read by attached microscopes.

(2) *Colby's.* In this, by a simple mechanical arrangement, two rods of different expansions are made to present two minute points or dots at a certain constant distance.

(3) *Struve's.* In this, one end of the bar carries a contact lever, the lower arm of this lever terminates in a polished hemisphere, while the upper arm traverses a graduated arc. When the index points to a certain division on the graduated arc, the bar is known to be at its normal length, and its length is also known, and can be allowed for, when the index points to other divisions.

109. We shall now proceed to explain the principle of General Colby's beautiful contrivance, known as the "COMPENSATION APPARATUS," which he invented for the Irish Survey, and with which the Lough Foyle base was measured.

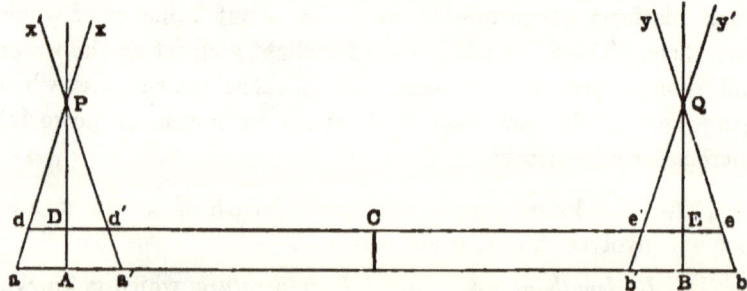

AB is a brass rod, DE is an iron rod: these are firmly united at their middle points by a transverse bar C, but their ends are free to expand or contract under changes in temperature. The

VII.]   BASE LINES.   125

lengths of these bars are the same at a certain temperature, but the expansion of the brass being greater than that of the iron, it follows that if the temperature is greater or less than that selected as the normal, the brass bar AB will expand to ab, or contract to $a'b'$, whereas under similar circumstances the iron bar DE will expand or contract to de or $d'e'$.

Now AP, BQ are two tongues of steel, about 6 inches long, attached to the extremities of the rods in such a way as not to interfere with the expansion or contraction of the two bars. A minute dot of platinum, almost invisible to the naked eye, is placed at P and Q on the tongues. Under all changes of temperature the distance between P and Q is exactly 10 feet.

At the normal temperature (60° F.) the two bars being of exactly the same length, we must have the tongue AP parallel to the tongue BQ, and perpendicular to the bars. Suppose the temperature to exceed the normal, then the bar AB expands to ab, but DE expands only to de, and therefore the tongue AP assumes the position ax, and BQ assumes the position by.

If therefore P and Q are taken at such a distance from the bars that $\dfrac{PA}{PD} = \dfrac{\text{expansion of } AB}{\text{expansion of } DE}$, it is evident that in the new positions of ax and by the distance between the points P and Q will not change, although their position above the bars will be somewhat lower. Suppose the distance between the two bars to be 2 inches, and let the expansion of brass : expansion of iron $= 83 : 53$, and let $PD = x$ inches, then we have

$$\frac{x+2}{x} = \frac{83}{53}, \therefore 53x + 106 = 83x, \therefore 30x = 106,$$

$\therefore x = 3\frac{1}{2}$ inches nearly.

110. About five or six sets of these rods are generally required in measuring a base.

The method of using the instrument is as follows. The rods are carefully levelled, and are placed so that the dot at the end of one bar is always at a fixed distance from the dot at the end of the next. This is secured by means of powerful microscopes known as "Compensation Microscopes," and constructed on the same principle as the "Apparatus" itself. These dots are brought under the microscopes, and are thus known to be 6 inches apart.

111. In 1845 Professor Bach invented for the United States Coast Survey a measuring apparatus combining the principles of Borda's rods, the compensation tongue of Colby's instrument, and the contact lever of Struve. Surprising results, both as regards time and accuracy, have been obtained by this "Compensating Base-Measuring Apparatus." More than a mile has been measured in a day; and Jeffers, in his *Nautical Surveying* (p. 105), states that by its means " a base of six miles in length has been measured with a probable error on remeasurement of one-tenth of an inch; which surpasses the accuracy of angular measurement with our present instruments." It may be remarked that the Lough Foyle base was measured at the rate of only 250 feet a day.

112. As a preliminary operation to the measurement of the base, it is usual after getting an accurate section of the line by the spirit-level to measure the distance in an approximate manner. A more detailed aligning follows. By the aid of the theodolite over the ends and intermediate points, pickets are driven into the ground at regular intervals; each picket carries a fine mark indicating exactly the line of measurement. This aligning requires all the care that can be bestowed upon it.

113. RATIO OF THE LENGTH OF THE BASE TO THE EXTENT OF THE SURVEY.

This ratio has varied much. Six bases have been measured in the United Kingdom, and these vary from five miles to over ten miles in length, the extent of the survey from north to south being over 750 miles. The base line at the Cape of Good Hope is eight miles long, the ground surveyed being about 600 miles from east to west, by 300 from north to south. In Syria about 200 miles have been surveyed, the two bases being each about a mile long.

The degree of *accuracy* required in a base line must be settled by the extent of the ground to be surveyed, and also by the object of the survey.

114. As just stated, *six* bases have been measured in the Ordnance Survey of the United Kingdom. The earlier by Ramsden's steel chains, the two most recent by General Colby's apparatus described above. The following table contains the measured lengths of the bases and their lengths in the corrected Triangulation, i.e. when regarded as "Bases of Verification."

Base Lines of the Ordnance Survey.

	Date.	Place.	Measured feet.	Computed feet.	Difference in feet.	County.
1.	1791,	Hounslow Heath,	27406·19,	27406·36,	+0·17,	Middlesex.
2.	1794,	Salisbury Plain,	36576·83,	36577·66,	+0·83,	Wilts.
3.	1801,	Misterton Carr,	26344·06,	26343·87,	−0·19,	Lincoln.
4.	1806,	Rhuddlan Marsh,	24516·00,	24517·60,	+1·60,	Flint.
5.	1817,	Belhelvie,	26517·53,	26517·77,	+0·24,	Aberdeen.
6.	1827,	Lough Foyle,	41640·89,	41641·10,	+0·21,	Londonderry.
7.	1849,	Salisbury Plain,	36577·86,	36577·66,	−0·20,	Wilts.

Clarke's *Geodesy*, p. 244.

About two miles were subsequently added to the Lough Foyle Base, making it the longest in the kingdom.

In the above list, numbers 1, 3, 5 were measured by Ramsden's steel chains: the differences between the measured and computed lengths in these cases are not greater than in the case of numbers 6 and 7, which were measured by the compensation apparatus. Hence it has been considered by some authorities that bases measured very carefully with steel chains are deserving of most confidence, and they have the further recommendation of being simple, cheap, and portable.

115. The lengths of bases have varied much in different countries. In East Prussia a line measured by Bessel was little more than a mile long, whereas in France there is one 11·8 miles. Between these as limits we have them of all lengths: Lough Foyle is over 10 miles; in India, except the line at Cape Comorin, 1·7 miles; *nine* others lie between 6·4 and 7·8 miles. In the Spanish Triangulation now in progress there are several about 1½ miles; the principal base near Madrid is 9·1 miles, and in the island of Iviça is one just a mile in length.

128   MARINE SURVEYING.   [CHAP.

It is sometimes customary to break up a line into segments and verify the lengths of these segments by triangulation, in other words to treat them as bases of verification. Thus, in the case of the Madrid base mentioned above, the total length is composed of five segments. Angles were observed at ten different stations, and then on the assumption that one segment is correctly measured, the lengths of the other four were computed. Regarding the middle segment as the basis of calculation, the measured and computed segments stand as follows:

Segment.	Measured Metres.	Computed Metres.	Differences Metres.
1	3077·459	3077·462	+ 0·003
2	2216·397	2216·399	+ 0·002
3	2766·604	—	—
4	2723·425	2723·422	− 0·003
5	3879·000	3879·002	+ 0·002
Sum	14662·885	14662·889	+ 0·004

Clarke's *Geodesy*, p. 173.

116. To reduce a Base to the Sea-level.

The base ought to be reduced to the mean level of the sea. This may be effected as follows:

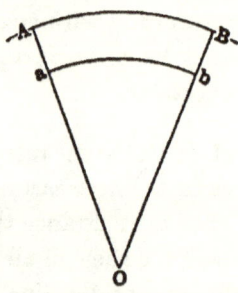

Let AB = measured length (A),
    ab = reduced length (a).
    r = radius of the earth, supposed to be a sphere,
    $r + h$ = radius at the mean level of the measured base.

BASE LINES.

Then by Euclid VI. 33, $\dfrac{a}{A} = \dfrac{r}{r+h}$.

$$\therefore a = \dfrac{Ar}{r+h} = A\left(1 + \dfrac{h}{r}\right)^{-1}$$

$$= A\left(1 - \dfrac{h}{r} + \dfrac{h^2}{r^2} - \dfrac{h^3}{r^3} + \ldots\right)$$

$$\therefore a = A - \dfrac{Ah}{r} + \dfrac{Ah^2}{r^2} - \dfrac{Ah^3}{r^3} + \ldots$$

$$\therefore A - a = \dfrac{Ah}{r} - \dfrac{Ah^2}{r^2} + \dfrac{Ah^3}{r^3} - \ldots$$

The first term only is required.

Example. In 1815 a base was measured in India equal to 30809·07 feet. The mean height of the base above the sea level was 1957 feet, and the mean radius of the earth is 20888153 feet.

Here
$\qquad A = 30809\cdot07$ feet,
$\qquad h = 1957$ feet,
$\qquad r = 20888153$ feet,

$$\therefore A - a = \dfrac{Ah}{r};$$
$\log A = 4\cdot488687$
$\log h = 3\cdot291591$
$\overline{7\cdot780278}$
$\log r = 7\cdot319900$
$\overline{\log(A - a) = 0\cdot460378} \therefore A - a = 2\cdot89$ feet.

$\therefore$ Reduced length of base = 30806·18 feet*.

Note. In ordinary marine surveying, such as the survey of a port, the surface of the earth may be considered as plane without sensible error, the arc of a degree not exceeding its chord by more than 25 feet.

117. Having thus noticed some of the more important methods of measuring a base line, we now pass on to the methods chiefly employed by the Nautical Surveyor.

---

* *Manual of Surveying for India,* p. 471.

In every base there are three elements:

(1) *The latitude and longitude of one extremity.*
This point is usually known as the Observatory Station.
(2) *The measured length.*
(3) *The direction or bearing of the base.*
We shall speak of these separately.

118. (1) THE LATITUDE AND LONGITUDE OF THE OBSERVATORY STATION.

The *Latitude* is usually determined by means of the sextant and artificial horizon, (1) By meridian altitude of the sun,
(2) By meridian altitude of many stars.
But in more exact surveys a zenith sector is required; by its aid the latitude can be determined to 1".

The *Longitude* must be determined very carefully by chronometers.

119. (2) THE LENGTH OF THE BASE.

Circumstances must decide the approximate length required: if the survey is limited, perhaps 1000 yards will suffice, but as a rule 2500 or 3000 yards will be deemed long enough.

120. *Nature of the ground.* The following conditions ought to be sought for as far as possible:

(α) As level as possible.
(β) Near the shore.
(γ) Not much above the sea level.
(δ) A good extent of country or shore, or many points in the harbour, to be visible from the proposed extremities of the line.
(ε) The extremities in view from one another.
(ζ). The line to run on to some well-defined natural object.

121. DIFFERENT METHODS OF MEASURING A BASE.

(α) By *masthead angle.*
(β) By *the velocity of sound.*

(γ) By *patent log under steam*.

(δ) By *astronomical observations*.

(ε) By *direct measurement*.

We will notice these methods one by one.

122. First Method. By Masthead Angle.

The distance of a ship from an observer on shore may be required, or from a boat, &c. The height of the mast is supposed to be known; a white mark (if required) may be placed on the side of the ship at the height of the observer's eye, and finally in observing the angle, the precaution must always be taken of measuring the angle "*on and off*" the arc, to eliminate the index error.

Then, *Length of base* = *height* × *cot angle*.

123. Second Method. By the Velocity of Sound.

Note. This method might be used under two circumstances:

(a) Where the required base would lie across a marsh, or bog, or the mouth of a creek.

(β) Where the base must be measured along coarse shingle.

Objections to this method.

(1) An error of $\frac{1}{10}$ second in the time produces an error of about 100 feet in the length of the base.

(2) The velocity of sound is affected by several variable conditions, viz. temperature, wind, and state of the atmosphere.

The mean of many observations will however give very good results.

Method of proceeding. A small gun is placed at one end of the proposed line, and a flag is hoisted to shew that everything is ready at the firing station. When this is answered from the observing station, the gun is fired, and the interval between the flash and the report is accurately noted. This operation is repeated several times, and at each station. The mean of each series is then taken, and if there is much difference in the results, the error is probably owing to the wind. If, however, the mean results are very close, their mean may be considered correct. The temperature must be also noted at each station.

Then the distance is computed from the formula,

Distance = time × velocity of sound at the given temperature.

The velocity of sound is 1090 feet per second at 32° F., and increases 10 feet per second for an increase of 9° F. in temperature.

Thus at 41° the velocity is 1100,
at 50° 1110,
at 59° .... 1120, &c.*

If the velocity of the wind must be taken into account, we proceed thus :—

Let D = required distance,

s = velocity of sound, corrected for temperature,

w = estimated velocity of the wind, (the wind is supposed to be in the direction of the measured base),

t_1 = time in seconds *with* the wind,

t_2 = time in seconds *against* the wind,

T = true time in seconds.

By 1st series of observations $D = (s+w)t_1$, $\therefore \dfrac{D}{t_1} = s+w$.

„ 2nd „ „ $D = (s-w)t_2$, $\therefore \dfrac{D}{t_2} = s-w$.

$$\therefore D\left(\frac{1}{t_1} + \frac{1}{t_2}\right) = 2s,$$

$$\therefore D = \frac{2s\, t_1 t_2}{t_1 + t_2};$$

but $\qquad\qquad\qquad D = sT.$

---

* The velocity may be more exactly computed from the formula
$V = 1093\sqrt{1 + \cdot 003665 t}$ where 1093 = velocity at 0°C,
t = observed temperature C.

$$\therefore sT = \frac{2s\, t_1 t_2}{t_1 + t_2},$$

$$\therefore T = \frac{2 t_1 t_2}{t_1 + t_2}.$$

Hence the true interval of time is the *harmonic mean* between the times with and against the wind. The true time being known, the length of the base can be found as above.

124. THIRD METHOD. BY PATENT LOG UNDER STEAM.

This can of course give but an approximate result; it is useful however in the case of a running survey of a coast, and will give a distance sufficiently correct for such a purpose, provided that, (1) the speed of the ship exceeds about three knots, and (2) the course is free from currents.

125. FOURTH METHOD. BY ASTRONOMICAL OBSERVATIONS.

This method also will only give a fair approximation, and will serve only in the absence of other methods, and for a rough survey of a large extent of coast. It is impossible to fix the position *exactly* by means of a sextant. This consideration will explain why the length of a base line is not determined on shore by finding the latitude and longitude of each extremity, and then computing the distance. In the first place, an error of 10″ in the latitude would involve an error of 1000 feet in the distance; and in the second place, the labour is very great to determine with great exactness the position on shore by accurate observations, such as would be necessary for the purpose.

126. FIFTH METHOD. BY DIRECT MEASUREMENT.

We now come to speak of the usual method resorted to in the case of most nautical surveys.

(a) *A sea base* can be measured as follows: When the surface of the water is free from currents and is calm, the distance between two boats, or between two rocks, can be measured by a line *well wetted* stretched between the points and supported by little floats at suitable intervals. The line should be measured at once in its wet state.

(β)   *A land base.*

This, being the most important, must be described in detail. When the ground has been selected combining as many of the advantages above mentioned as possible, the ends of the intended line (AB) are marked with easily distinguishable objects. A theodolite is fixed immediately over one extremity A, the stake having been removed, and the alignment is then effected as follows. Staves are placed, at intervals of 300 or 400 feet, in a direct line with station B, the true line being ascertained by the telescope of the theodolite. When these staves are firmly driven in, a deep-sea sounding line is stretched taut along the ground so as to touch all the staves on the same side. We thus obtain a straight line between A and B. The actual process of measuring is then carried on.

One man, called the Leader, being furnished with ten iron pickets about 12 or 14 inches long, and sharply pointed, takes the surveying chain (*vide* description in the preceding chapter) by one handle, and moves off from A towards B. When the chain has been stretched tight along the ground by the side of the lead line, one end is held over the hole at A by a second man, known as the Follower, and the Leader drives in a picket at the further end. The chainmen then move on; the Follower comes to the picket, and holding his end close up to it the Leader drives in a second picket. The Follower then makes an entry in his note-book that one chain has been measured. They proceed thus, the Follower retaining all the pickets he has removed, until the Leader has exhausted his ten pickets, when the Follower transfers the nine pickets in his possession. When this happens ten chains have been measured, and the record in the Follower's book will stand thus ⊮⊮. When station B is approached the links and inches are measured from the last picket in the ground, and a note of this interval made in the book*. Thus the total distance from A to B has been measured. The whole process of measuring is then repeated from B to A, and if the results are not very different the mean of the two results may be regarded as correct. If the measurement is carefully

* Technically known as the Field Book.

made the result ought to be less than a yard in error in a distance of 1000 or 1200 yards. The sources of error are twofold: (1) *Overriding of links*, which is generally due to carelessness. A third observer to walk along the chain when stretched will prevent this occurring. (2) *Unevenness of the ground.* Hence some practical men have recommended long narrow boards to be laid down by the line where the ground is very uneven.

127. (3) The Direction or Bearing of the Base.

This is the third and last "element" of the base line.

(α) From the Observatory Station A at one end of the line, obtain with the sextant or theodolite the horizontal angle between the sun's N. L. and station B.

(β) Compute the sun's True Bearing, either by altitude or time azimuth*.

We can then compute the T. B. of the station B, and hence the direction or bearing of the base.

Example 1. The horizontal angle between the second extremity of the base and the sun's centre was found to be $114° 26' 00''$; sun to the right of the object: the sun's T. B. was found to be N. $75° 58'$ E.: find the direction of the base.

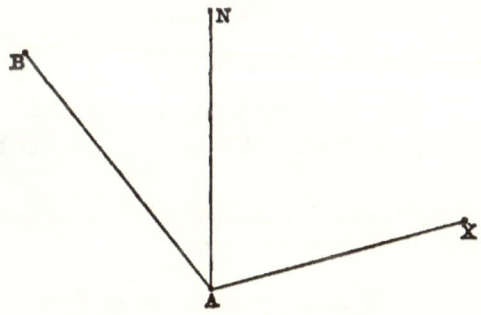

T. B. = N. $75°-58'$ E. = NAX.

* The Sun's True Bearing at any hour can be found with great facility by means of the "Sun's True Bearing or Azimuth Tables computed for intervals of four minutes." These tables are supplied to ships by the Admiralty.

Observed angle = 114°-26' = BAX.
∴ NAB = 38°-28',
∴ Direction of the base $AB = N\,38°-28'\,W$.

Example 2. The horizontal angle between an object O near the horizon to the left of the sun and the sun's centre was computed to be 126°-15', the sun's T. B. was computed to be S. 67°-49' E., and the angle between the object O and station B was found by theodolite to be 12° 26', O being to the left of B: find the direction of the base.

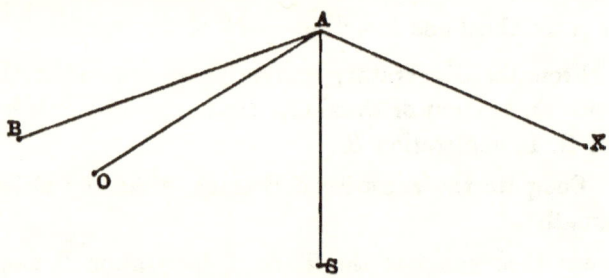

X represents the place of the sun.
Sun's T. B. = S. 67°-49' E. = SAX.
Obs. angle = 126 - 15 = OAX.
∴ 58 - 26 = SAO,
    12 - 26 = OAB.
∴ 70 - 52 = SAB,
∴ Direction of the base line AB is S. 70° 52' W.

128. *Note.* The angle between the sun and the object is seldom horizontal; the horizontal angle may however be computed from the formula*.

Log. Cos Hor. Angle = log Cos obsd. distance to sun's centre
           − log Cos sun's app. alt.

129. In selecting an object to be observed with the sun, the following conditions ought to be sought for as far as possible. The object ought to be (1) well defined, (2) low, (3) near the horizon, and (4) not less than 90° from the sun.

* Vide Todhunter's *Spherical Trigonometry*, p. 92. 3rd Edition.

Examination.

(1) What is the use of a base line in a survey?

(2) Define a "base of verification."

(3) Specify the various methods by which the length of a base may be determined.

(4) In finding a base by masthead angle, what precaution must be taken?

(5) Distinguish between standards "*à traits*" and standards "*à bouts.*"

(6) "The geodetic standards vary in different countries;" what does this mean?

(7) Mention the causes of variation in the lengths of measuring rods.

(8) What contrivance has been found to insure the uniformity of length between two points in a measuring rod? Draw a diagram, and fully explain the principle.

(9) Describe fully the method by which a base line is measured with the surveying chain.

(10) Specify the most favourable conditions to be sought in the ground for a base line.

(11) What are the "three elements of the base"?

(12) How are the Latitude and Longitude of the Observatory Station usually determined in a marine survey?

(13) How is the Direction of the base line found?

(14) Why is the direction of the base important?

(15) Under what circumstances may a base measured by patent log be of use?

(16) Explain the method of finding the Length of a base by the velocity of sound.

(17) In the case of wind blowing in the direction of the base, when the distance is to be measured by sound, investigate a formula by which the true distance may be computed.

(18) If the time *with* the wind is 16 seconds, and *against* it 18·5 seconds, calculate the true time; and if the temperature is 62° F., compute the length of the base.

Results. Time = 17·16 seconds,

Distance = 20462 feet.

(19) The mean of a set of observations taken at Pile △, lat. 52° 00′ N., was as follows :—

Zero—Steeple—360°. 0′—Mag. S. 32° W.

At $5^h\ 16^m$ P.M. app. time 85° 00′ ☉

Find the true bearing of Steeple from the Pile, and the variation.   Sun's decl. 20° 30′ N.   (June, 1877.)

(20) The mean of a set of obs. taken at theodolite △ X, in lat. 53° 14′ N. was as follows :—Sun's decl. = 16° 38′ N.   Sun's semi. = 15′ 50″.

Zero—Camp △ —360° 0′—Mag. N. 45° E.

At $6^h\ 0^m$ A.M. app. time 37° 25′ ☉

Required the true bearing of the Camp △ from X, and the variation.   (April, 1876.)

(21) The mean of a set of obs. taken at Cairn △, in lat. 50° N., was as follows :—Sun's decl. 12° 30′ N.

Zero—Wedge △—360° 0′.—Mag. N. 30° E.

$6^h\ 0^m$ A.M. app. time 40° 3′ ☉

Required the true bearing of the Wedge from the Cairn, and also the variation.   (Oct. 1876.)

(22) At theodolite △ X, in lat. 54° 3′ N., the following observations were taken.

Zero—Lighthouse (southward of X)......360° 0'.

First set with sun's *lower* and *right* limbs.

$3^h\ 58^m$ P.M. 35° 11' alt. ...... 104° 19'
35° 6' „ ...... 104° 26'
35° 1' „ ...... 104° 33'.

Second set with sun's *upper* and *left* limbs.

$4^h\ 0^m$ P.M. 35° 30' alt. ...... 104° 4'
35° 26' „ ...... 104° 11'
35° 21' „ ...... 104° 18'.

Sun's declination = 19° 36' 6" N.

Required the true bearing of the Lighthouse from X.

(Aug. 1875.)

Answers to Examples 19—22.

(19) True Bearing = S. 9° 39' W.
      Variation = 22° 21' W.

(20) True Bearing = N. 42° 10' E.
      Variation = 2° 50' W.

(21) True Bearing = N. 41° 50' E.
      Variation = 11° 50' E.

(22) True Bearing = S. 30° 17' E.

CHAPTER VIII.

TRIANGULATION.

130. WHEN the base line has been measured the triangulation of the survey may be proceeded with. It has been already mentioned that if we know the length of one side of a triangle, and the values of two of its angles, the lengths of the other sides may be computed.

If a conspicuous object, which we will denote by C, be visible from the extremities A and B of the base, the angles BAC, ABC are observed with the theodolite or sextant, and the line AB being known, we can compute the sides AC and BC.

Note. If at C the angle ACB be also observed, and the sum of the three observed angles is exactly 180°, the triangle is technically said to "close"; but this is very seldom the case: the error may be due to the imperfection of even the very best instruments, or to the "Personal Equation" of the observer.

131. It is probable that no two observers will ever see the same phenomenon at *exactly the same moment of time*, e.g. the instant of a star's transit over a wire. One will see the event a very little before or a very little after another. The difference between the time of observation of some one person, who is known as the Standard, and the observer's time, is called the *Personal Equation* of the observer. Similarly in taking angles with a large theodolite in an important survey, the Personal Equation has to be taken account of.

Thus in the "*Report of the Difference of Longitude between the United States Naval Observatory and the Sayre Observatory of Lehigh University*" we have this notification: "My Personal Equation as determined by the use of the instrument at the Washington Observatory was $0^s\cdot 126$; the transit observed too soon*."

132. The selection of the stations for the Triangulation requires considerable experience, as much depends on the nature of the district and the object with which the survey is undertaken.

The triangles should be as nearly equilateral as possible, because when the angles are all equal, an error in one of the observed angles will produce a minimum of error in the computed sides. If the equilateral triangles are not obtainable, then they must be as "*well-conditioned*" as possible, i.e. the angles must lie between 30° and 75°. An "*ill-conditioned*" triangle may be defined as one in which the two sides are very long when compared with the base, and therefore the base angles are very much larger than the vertical. It is equally advisable to avoid the contrary error, of having too large a vertical angle.

133. *Increasing from the Base.*

The base line being usually very small in an extensive survey, compared with the distances between the principal points of the triangulation to be ultimately derived from it, the sides of these triangles must as quickly as possible be increased, until they arrive at their greatest limit, and this limit is simply the distance at which they are mutually visible. The following diagram will explain the method of increasing the lengths of the "Primary Triangles" directly from the measured base.

Let AB be the measured base, C and D represent two trigonometrical points. Observe all the angles, viz. BAC, ABC, BAD,

---

* In this Report, which may be seen in the Library of the Royal Naval College, there is a description of a "Personal Equation Apparatus," designed by Prof. Eastman, for ascertaining the Personal Equations of Observers.

ABD, and compute the distances of *C* and *D* from *A* and *B*; then in the triangles *DAC*, *DBC*, we have two sides and the included

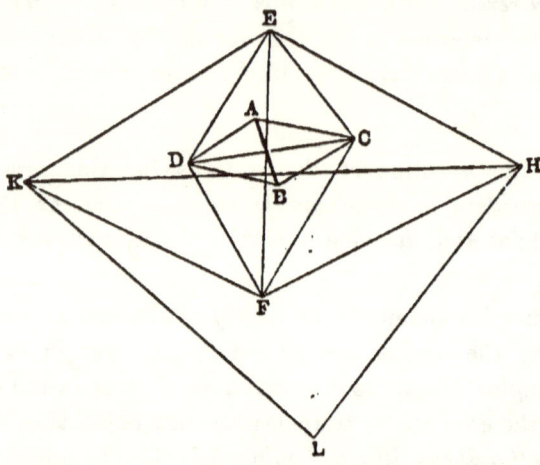

angle, and hence *DC* can be computed from each; we thus have a check upon the accuracy. Next, the distances of *E* and *F* from *C* and *D* are calculated, and, as before, *EF* is found from two triangles. In the same manner the length of *KH* is found; and this method is carried on until the trigonometrical stations are at the required distance apart. When this has been accomplished, the primary triangles are piled one upon another until the whole country is embraced; arrangements are also made either by a second chain of triangles, or else by the triangles overlapping, so that independent values may be obtained for the length of the sides. This is known as the Principal Triangulation. When completed, these *Primary* triangles are divided into smaller triangles, called *Secondary* triangles; these are again subdivided into others, designated as *Tertiary* triangles, and so on, until the sides of the triangles are about 2 miles long, when they may be considered as straight lines without sensible error.

134. In the Ordnance Survey of Great Britain the largest-sized theodolites, 3 feet in diameter, were used in fixing the principal stations. The angles of the secondary triangles were observed with smaller instruments. In the Primary triangulation there are 218 stations, at 16 of which no observations were made. The number of observed bearings was 1554. The longest side of any primary triangle was upwards of 111 miles, and connected Slieve Donard in Co. Down with Sca Fell in Cumberland. The sides of the secondary triangles average about 8 or 10 miles, and those of the tertiary from 1 to 3 miles. Theodolites of 7, 9, and 10 inches diameter were used in measuring the angles of these secondary and tertiary triangles. Ramsden was the maker of the great 3-feet theodolite.

135. In the First Order of triangles only the most remarkable objects in the country are noted. On these objects signals are placed, and the distances are computed from the Measured Base Line. In the Second Order a series of prominent points are noted in each Primary triangle; these points form triangles, *one of which has a side in common with the Primary triangle*, and hence is accurately known. In the same way a Tertiary triangle has a side in common with a secondary triangle, and thus a rigorous exactitude is kept up.

136. The length of the sides of the smaller triangle depends on the minuteness of the survey; e.g. if the contents of parishes, estates, &c. are to be calculated, then the sides ought not to be more than 1 or $1\frac{1}{2}$ miles for an enclosed country like England, and 2 or 3 miles for a more open country. If no contents are required, and the triangulation is merely to correct a topographical survey, then the distances will depend on the scale of the map. In the Ordnance Survey maps constructed on the scale of 6 inches to a mile, *two* points per square mile were fixed during the triangulation, while on the maps constructed on the scale of 60 inches to a mile, *sixteen* points per square mile were determined.

137. The reason is, that under the most favourable conditions for chaining, a distance so measured is likely to have 20 times the amount of error it would have if determined by triangulation, and hence the larger the scale on which the work is plotted the closer must be the trigonometrical points if perceptible error is to be avoided.

138. Very distant stations are generally observed at night, the atmosphere being then more adapted for delicate observations. The stations are generally observed by means of Drummond's Light* fixed on them. This consists of a ball of lime $1\frac{1}{4}$ inches in diameter placed in the focus of a parabolic reflector and raised to an intense heat by a stream of oxygen gas, directed on it through a flame of alcohol. The light thus produced is 80 times as intense as that given by an Argand burner. An example of its great power was afforded in the Irish Survey, in which an important station could not be observed from Devis Mountain, near Belfast, until this light was erected, when, notwithstanding most unfavourable conditions in the weather, the light was brilliantly visible at a distance of nearly 70 miles, and could have been observed at a much greater distance.

139. ASSUMED BASE.

Sometimes the triangulation may be carried on before the Base is measured, and then the calculated sides corrected when the Base has been found; or, if an error has been made in measuring the Base, the sides of the triangles computed from the erroneous length can be corrected afterwards.

Thus, let AB = erroneous Base

AD = true Base.

Let BC, AC = the distances to the trigonometrical point E calculated from the erroneous Base.

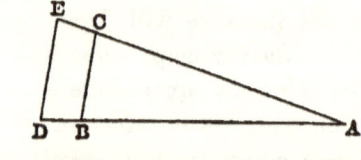

By Euclid vi. 4, $\dfrac{DE}{BC} = \dfrac{DA}{BA} \therefore DE = BC \cdot \dfrac{DA}{BA}$

and $\dfrac{EA}{CA} = \dfrac{DA}{BA} \therefore EA = CA \cdot \dfrac{DA}{BA}$

and DE, AE are the true distances required.

* So called from Lieut. Drummond, who employed it in 1826.

140. Spherical Excess. All the observed angles are essentially the angles of spherical triangles, and the three angles of every spherical triangle are together greater than 180° (vide Todhunter's *Spher. Trig.* p. 13). The lines containing the observed angles are really tangents to the earth's surface, whereas to obtain the three points considered as the angular points of a plane triangle, the observed angles must be reduced to the angles contained by the chords of the arcs which form the spherical triangle (Todh. p. 73). The correction for the "*spherical excess*" is much too small to be applied to angles observed with moderate sized instruments, being completely lost in the greater errors of observation. It must however be taken account of in the primary triangulation. Thus in one of the large triangles of the Ordnance Survey the sum of the three angles was $0''\cdot 5$ less than 180°, and the calculated spherical excess amounted to $1''\cdot 29$, shewing an error of $1''\cdot 79$ in the observations. The practical rule adopted is to add *one-third of the error* to each of the observed angles and thus find the angles of the spherical triangle, and then subtract one-third of the spherical excess from each of the corrected angles, and thus obtain the angles of a plane triangle ready for calculation.

Note. The Spherical Excess is obtained from the formula

$$A + B + C - 180 = \frac{S}{r^2},$$

or, the excess in seconds $= \dfrac{S}{r^2 \sin 1''}$, where S = area of the triangle, and r = radius of the earth*.

141. The Reduction to the Centre. This is the term applied to the correction when the theodolite is placed near, but not exactly over, the point denoting the station.

---

* Vide Todhunter's *Spher. Trig.* Chapter IX. throughout; and also Snowball's *Spher. Trig.* Chapter v., especially §§ 62—73. The General Rule stated above is derived from Legendre's Theorem, viz.:—"If each of the angles of a spherical triangle, whose sides are small when compared with the radius of the sphere, be diminished by one-third of the spherical excess, the triangle may be solved as a plane triangle whose sides are equal to the sides of the spherical triangle, and whose angles are those reduced angles."

Suppose A to be the station, and the angle CBD to have been observed from the position B, then the angle CAD is required.

Observe CBD, CBA, and measure AB.

$$COD = CAD + ACB,$$
$$COD = CBD + ADB;$$
$$\therefore CAD + ACB = CBD + ADB,$$
$$\therefore CAD = CBD + ADB - ACB.$$

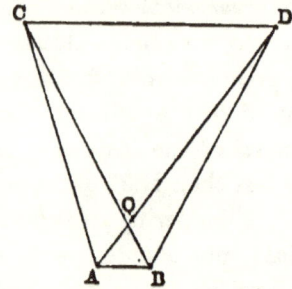

But, $\dfrac{\sin BDA}{\sin DBA} = \dfrac{AB}{AD}, \therefore \sin BDA = \dfrac{AB}{AD} \cdot \sin DBA$;

also, $\dfrac{\sin ACB}{\sin ABC} = \dfrac{AB}{AC}, \therefore \sin ACB = \dfrac{AB}{AC} \cdot \sin ABC.$

These angles being very small, we have

$$CAD = CBD + \dfrac{AB}{AD} \cdot \sin DBA - \dfrac{AB}{AC} \cdot \sin ABC\ *.$$

This correction is not often required, as in the primary triangulation care is taken that only those stations are selected which can be actually observed from, and in the secondary and tertiary triangles it is not necessary to *observe* the third angle.

142. The triangulation of this country was connected in 1861 with that of France and Belgium. The necessary operations were carried on by English officers acting in concert with officers of the other countries. The work was commenced in June and was finished in the following January. The following instances

* The values of AD and AC are known very approximately from the side CD and the observed angles CDA, DCA.

VIII.]   TRIANGULATION.   147

of accuracy in the work carried on independently are interesting. The sides of certain triangles were computed by the English and Foreign officers, working from their own bases, the English from the British Triangulation carried across the Channel, and the Belgians from their measured base at Ostend.

Distance.	Miles.	English Result in metres.	Belgian Result in metres.	Difference.
Hondschoote Kemmel	17·2	27612·80	27612·74	− 0·06
Cassel Hondschoote	13·3	21415·30	21415·34	+ 0·04

143. Colonel Clarke gives the following interesting account of the operations by which the triangulation of Europe was connected with that of Algiers and North Africa*. In 1868 M. Perrier (one of the French officers employed in 1861 in connecting the triangulation of Great Britain with that of the Continent) satisfied himself that it was possible to connect geodetically Algiers with the peaks of the Sierra Nevada, distant 60 leagues away in Andalusia. In 1879 the French and Spanish officers succeeded in accomplishing the work.

Twenty miles south-east of Grenada is the highest peak in Spain, Mulhacen, 11420 feet. Distant 50 miles ENE. from this is Tetica, 6820 feet. The line joining these points forms one side of a great quadrilateral; the opposite side being in Algiers. The terminal points of the African side, which is 66 miles long, are Filhaousan (3720) and M. Sabiha (1920 feet), each of these mountains being about 170 miles from Mulhacen. The other two sides and the diagonals of the quadrilateral span the Mediterranean. At each station the signal light was produced by a steam-engine of 6 H. P., working a Gramme's, magneto-electric machine† in connection with the apparatus of M. Serrin. The labour of transporting to such altitudes this machinery, with the requisite water and fuel, in addition to the ordinary geodetic instruments and equipment, and the maintenance of the whole in

* *Geodesy*, p. 261.
† Vide Ganot's *Physics*, p. 791.

a state of efficiency for two months, necessitated the formation of a military post at each station. After immense difficulties the whole was ready on August 20. It was not, however, until September 9th that the electric light on Tetica was seen in Algiers; a red round star-like disk visible at times to the naked eye. On the following day Mulhacen was seen, and the observations were carried on until October 18th, with results which leave nothing to be desired in point of precision. Thus a continuous triangulation now extends from the Shetland Isles into Africa.

Examination.

(1) What is meant by the Triangulation of a country?

(2) Define a Primary, Secondary, and Tertiary, Triangulation.

(3) What is meant by the expression "the triangle closes"?

(4) What two causes tend to prevent a triangle closing?

(5) What is the "Personal Equation" of an observer?

(6) What is an "ill-conditioned" triangle?

(7) What conditions ought to be sought in selecting a triangle for surveying purposes?

(8) Define "Spherical Excess;" and explain how it is allowed for in computing the triangles of a Trigonometrical Survey.

(9) Write a note on the method of reducing the observed angles of a primary triangle to a form for computation.

(10) Define "Reduction to the Centre;" investigate the value of the correction; and draw an explanatory diagram.

(11) Explain the necessity of an accurate base line, and shew how an error in the measured base may be afterwards allowed for.

(12) In triangulating between distant objects, explain the cause which renders it necessary that a correction should be applied to the observed horizontal angles for spherical excess.

(Sept. 1879.)

CHAPTER IX.

LEVELLING.

Note. Before studying this chapter the student ought to read the description of the Spirit Level in Chapter VI. (§ 88).

144. DEF. Several points are said to be *on a level* when they are on the same surface concentric with the surface of still water.

DEF. Every line traced on the surface of the earth, assumed to be a sphere, is called a *Line of True Level.*

DEF. Every line in a plane perpendicular to the direction of a plumb line is called a *Line of Apparent Level.*

DEF. The difference of level of two points is the difference of the radii of the spheroid upon which the points are situated.

The object of levelling is to ascertain the relative heights of objects.

DEF. A continuous line passing through these objects, supposed to be in the same vertical plane, forms a *Section* or Profile.

145. The heights of these objects are estimated from an imaginary line, either above the highest point or below the lowest point (generally the latter), and this line is known as the *Datum Line.*

Thus in the Ordnance Survey of Great Britain and Ireland the Datum Line is the level of the mean tide at Liverpool. In Calcutta the levels of the city are all referred to the bottom, or sill of the stone at the tide gauge in the dockyard, and this point is 8·83 feet below the mean sea level. The points where the levelling staves stand when the levelling is in process are known as Bench Marks (B.M.), and are usually engraved on some permanent object, such as a mile stone, or curbstone of a bridge*. The symbol used in the Ordnance Survey is the well-known broad arrow (↑). These marks serve for future verification, and when the work is carried on from day to day. From these permanent positions side lines of levels can be taken at any time.

DEF. A *Check Level* is generally run to test the accuracy of the more detailed work. If the difference of level between the two extreme points is found to correspond with that previously ascertained, the presumption is that all the intermediate work is correct.

146. Levelling is of two kinds, *Simple* and *Compound*. *Simple Levelling* consists in finding the difference of height between two points A and B as follows: Suppose the instrument placed at any point and a levelling staff erected at A; the reading is noted; the staff is then removed to B, and the reading noted; then the difference of these readings will be the difference of the heights of the points A and B. E.g. At A the reading was 7·25 feet, and at B the reading was 13·64 feet, then we infer that the point B is 6·39 feet *below* A.

Compound Levelling is the term applied when a series of levels must be taken between two points far apart, and not mutually visible. Each of these levels is accomplished as follows:

147. The Spirit Level is placed *midway* between two levelling staves, but not at a greater distance than 250 feet from each. At greater distances the minute divisions on the staves cannot be

---

* One of these marks in the Royal Naval College is on the wall of the building near the door by which the room is reached where the Acting Sub-Lieutenants are examined, and another is near the entrance to the Office of the Director of Studies.

accurately distinguished even in the best telescopes (vide § 89). This method eliminates the errors arising from the curvature of the earth and from refraction. *The difference between the two readings* will then be equal to the difference of the heights of the two stations.

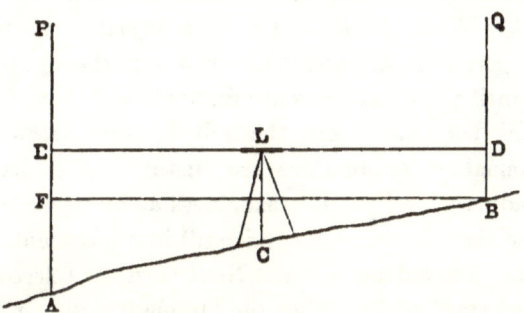

Let A and B be the two stations, the difference of whose heights is required. Let AP and BQ be two levelling staves held at A and B respectively. And let L be the Level erected at C midway between A and B.

Looking *back* the height AE is read off: looking *forward* the height BD is read off. Then by Euc. I. 34, we have $BD = EF$; $\therefore AE - BD = AE - FE = AF =$ difference of the heights.

Suppose the back reading AE to be 10·7 feet, and the fore reading BD to be 2·9 feet; then $AF = 7·8$ feet, the height of B above A.

If BD were the *back* reading, and AE the *forward* reading, then we would say that the point A was 7·8 feet below B.

Hence the rule:—When the back reading exceeds the fore reading, the process is being carried on *up* hill, but when the back reading is less than the fore reading, the work is going on *down* hill.

Example. The back reading is 12·52 feet, and the fore reading is 13·26 feet, which is the higher point?

Result. The back station is 0·74 feet above the fore station.

148. The following method is generally pursued in running a series of levels through a tract of country.

A staff-holder places his staff on the bench mark from which the levels are to commence. The Spirit-Level is set up in the most favourable place not more than about 250 feet from the staff, and in the direction in which the levelling is to be carried on; a second staff is held at the same distance in front of the instrument. When the Level has been adjusted according to the directions given in Chapter VI., draw out the eye-piece of the telescope until the cross-wires are distinct, and then, directing the glass to the back staff, turn the milled-headed screw on the side until the smallest graduations are distinct. The back reading is then made with all possible exactness and entered in the Field Book. See that the instrument is still in adjustment, and read a second time to avoid any error. Next turn the telescope towards the forward staff, and see that the bubble is still in the centre of its "run." Read the divisions on the fore staff most carefully, and note the result; then read a second time for verification. This completes the first levels.

The surveyor passing the forward staff-holder proceeds to some convenient spot to set up the instrument a second time, and this spot, as before remarked, ought not to be at a greater distance from the staff than 80 or 90 yards. The man who held the back staff now proceeds about the same distance in front of the instrument.

The instrument is again adjusted, and the second levels completed. In this way the work proceeds until the second terminal station is reached. A diagram will perhaps make it easier to understand the process.

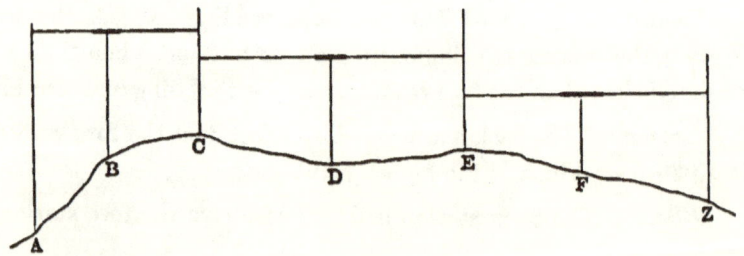

Suppose we wish to ascertain the difference of level between the two extreme points A and Z. The staves ought to be erected at every point where a serious irregularity exists in the ground. Let two staves be erected, one at A and the other at C. The Level is placed at B midway between them, and the readings on the staves at A and C are noted and entered. The Level is next placed at D, and the staff at C being simply reversed, the graduations being now towards Z, the staff at A is taken forwards to E. The readings at C and E are noted and entered as before. The instrument is then taken to F, and the staff at C to the point Z, and the readings are noted.

Now by what has gone before the difference between the readings at A and C is the difference in the height of A and C; similarly we can obtain the difference of the height of C and E, and of the height of E and Z, and thus we can find the difference between the heights of A and Z. In fact, the difference between the *sums* of the back and fore readings will be the difference between the heights of the initial and terminal stations, the higher of the two being determined by the usual rule (p. 151).

Example to illustrate this:

Stations.	Back Reading.	Fore Reading.
A and C	10·46 ft.	11·20 ft.
C and E	11·33	8·00
E and Z	7·42	7·91
Sums =	29·21	27·11
	27·11	
Difference =	2·10	

∴ A is 2·10 feet *lower* than Z.

149. To lay down a Section.

In running a *check level* we require to enter only the back and fore readings, and any remarks that may be necessary in a column reserved for the latter purpose in the Field Book. But to lay down a *section* the following form of the Field Book is necessary to register all the information required.

154   MARINE SURVEYING.   [CHAP.

1	2	3	4	5	6	7	8	9	10	11	
No. of Station.	Back Reading.	Back Bearing.	Back Distance.	Fore Bearing.	Fore Distance.	Fore Reading.	Rise.	Fall.	From Initial Station.		Remarks.
									Rise.	Fall.	
	Ft. Dec.		Ch. Lks.		Ch. Lks.	Ft. Dec.	Ft. Dec.	Ft. Dec.	Ft. Dec.	Ft. Dec.	
1	13·71	205°	5·19	25°	7·96	7·88	5·83	—	5·83	—	
2	9·40	208°	2·27	25°	3·08	16·30	—	6·90	—	1·07	
3	3·87	207°	5·08	23°	3·40	11·71	—	7·84	—	8·91	
4	2·63	208°	6·59	28°	4·00	12·41	—	9·78	—	18·69	
5	14·62	205°	3·92	26°	5·20	0·95	13·67	—	—	5·02	
6	17·00	208°	4·64	29°	3·89	1·45	15·55	—	10·53	—	
Sum of Back	61·23										
Sum of Fore	50·70				50·70						

Diff.=10·53 which corresponds with the result in Col. 10.

Hence we infer that Station No. 6 is 10·53 feet above Station No. 1.

In delineating a section on paper, especially if the irregularities are not very marked, it is necessary to exaggerate the vertical scale in order that these irregularities may become apparent. This process of course distorts the true appearance of the ground. The horizontal scale is usually made some exact part of the vertical, so that the proportion may be apparent to the eye. Thus for the *vertical* scale we may have 25, 50, 100, or 150 feet to an *inch*, according to the amount of detail required, and then for the horizontal scale we may take from $\frac{1}{2}$ to $\frac{1}{10}$ of the above, or even less, if the section is of great length, and the ground generally flat.

150. *Example.*

Let the following data be taken from the Field Book. Distances, 650, 700, 750, 670, 600, 650, 500, 750 feet, and the Differences, Fall 12·2, Fall 18·32, Fall 14·09, Fall 0·21, Rise 8·32, Fall 2·4, Fall 24·44, Fall 37·79 feet, reckoned from the initial

LEVELLING.

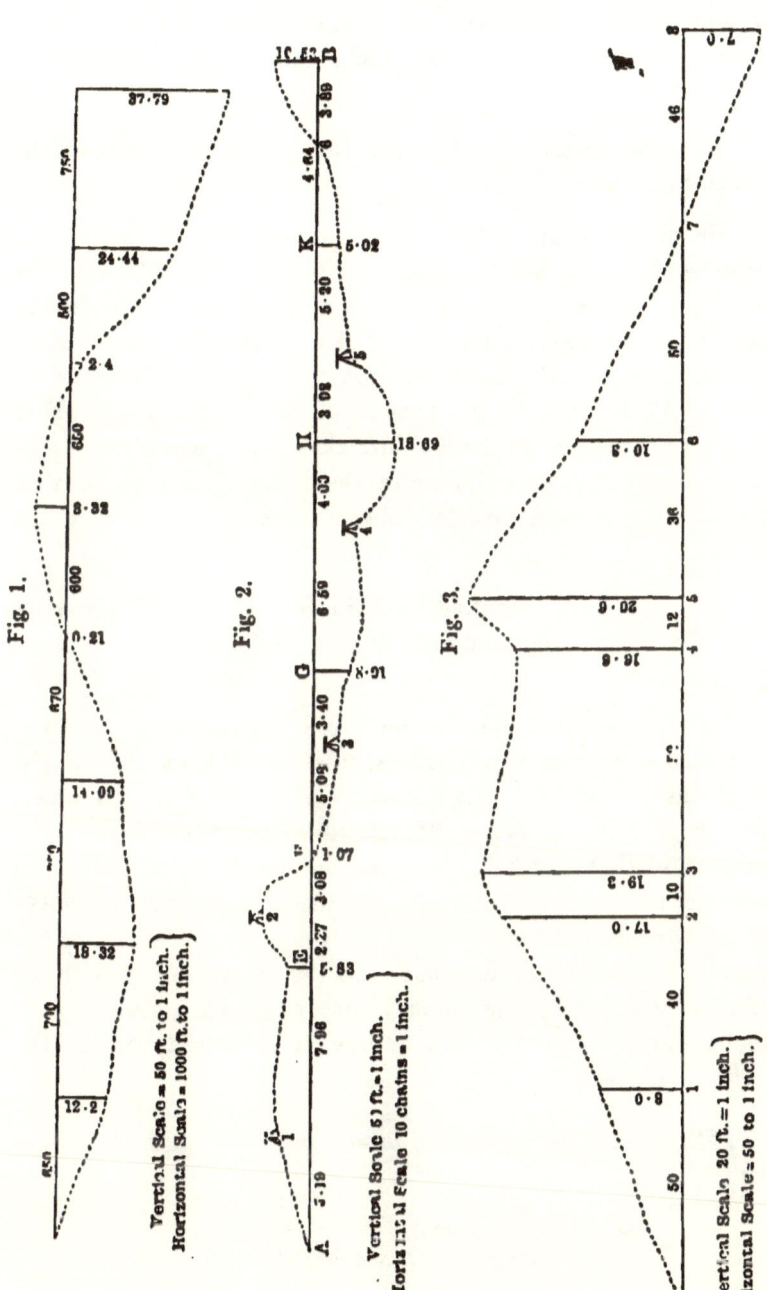

Fig. 1.
Fig. 2.
Fig. 3.

station. To form a section from these elements we may take a vertical scale of 50 feet to one inch, and a horizontal scale of 1000 feet to one inch.

The annexed diagram (fig. 1, p. 155) exhibits the section obtained from the above data *.

We simply draw a horizontal line, and lay off on it the distances 650, 700, 750 feet, &c. on the scale of 1000 feet to the inch, and at these points draw perpendiculars above or below the horizontal line according as the difference of height is a rise or fall from the initial station, and on these vertical lines lay off 12·2, 18·32, &c. feet, on the scale of 50 feet to the inch. Then, drawing a line through these points thus found, and attending to any remarks that may appear in the column reserved for that purpose in the Field Book, we obtain a section or profile of the ground.

151. Let it be required to delineate the section of ground from the information contained in the Field Book given above (§ 149) †.

Draw a horizontal line AB (fig. 2, p. 155), and lay off on it the back and fore distances from columns 4 and 6, on the scale of 10 chains to one inch. At the points E, F, G, H, &c. erect perpendiculars above or below AB, according as the ground at these stations is above or below the station A. This information is supplied in columns 10 and 11. On these perpendiculars we lay off the rise or fall thus obtained on the scale of 50 feet to the inch. Join the points thus found, having respect of course to the column for remarks (wherein may appear that the stations are on knolls, necks, &c.), and the outline thus formed will be the section required.

152. Let it be required to form a section from the following data.

---

* Frome, p. 98.
† *Manual of Surveying for India*, p. 547.

1	2	3	4	5	6	7	8	
No. of Station.	Back Reading.	Fore Reading.	Rise.	Fall.	Rise. From Initial Station.	Fall.	Distance in feet.	Remarks.
1	12·0	4·0	8·0	—	8·0	—	50	
2	11·5	2·5	9·0	—	17·0	—	40	
3	10·2	7·9	2·3	—	19·3	—	10	Knoll.
4	2·1	4·8	—	2·7	16·6	—	52	Neck.
5	7·4	3·4	4·0	—	20·6	—	12	Knoll.
6	2·6	12·9	—	10·3	10·3	—	36	
7	1·4	11·7	—	10·3	—	0·0	50	
8	3·6	10·6	—	7·0	—	7·0	46	Level of water.

In this case measure off the distances 50, 40, 10, &c. along the horizontal line (fig. 3, p. 155) on the scale of 50 feet to one inch; erect the perpendiculars at the points above the line or below it as required, and set off along these perpendiculars the lengths 8·0, 17·0, 19·3, &c. from columns 6 or 7, on the scale of 20 feet to one inch. Then, as already explained, the dotted line will represent the section required.

In addition to the method of finding the difference of height between two points, already explained, three other methods are sometimes employed.

(a) *By Theodolite.*
(b) *By Barometer.*
(c) *By Thermometer.*

153. BY THEODOLITE.

The instrument is set up at one extremity of the line to be levelled; and every irregularity of the ground being marked by pickets, a levelling staff, furnished with a vane which can be fixed at any required height, is set up at the first serious change in the ground. The vane being set at the exact height of the telescope, the angle of depression is taken. The instrument and the staff are then made to change places, and the angle of elevation is observed. The mean result may be considered as correct, and then the vertical arc being clamped to this angle the cross-wires are made to bisect the vane.

The annexed diagram will explain the method of proceeding*:

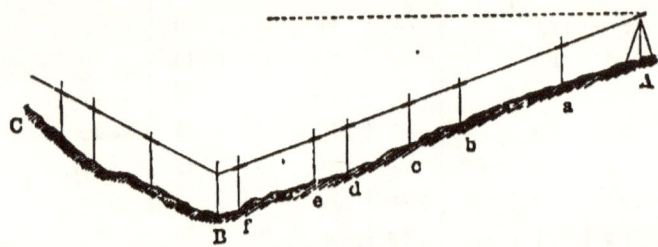

Let a Theodolite be set up at A, and a levelling staff, furnished with a sliding vane, be held at B. The vane is fixed at the exact height of the axis of the telescope. Between these two stations a series of intermediate positions a, b, c, &c., at which a levelling staff is to be erected, will be determined by the irregularity of the ground.

The angle of depression from A to B is observed, and then the cross-wires are made to bisect the vane at B. The vanes on the staves at a, b, c, &c. are then shifted until the line of sight passes through their centres. On arriving at B, after reading the heights of the vanes at a, b, c, &c., and measuring the distances Aa, ab, bc, &c., the Theodolite is placed at B, and the process is repeated up the hill to C; and thus the work goes on.

154. To lay down a section on paper from the data thus obtained, we must first draw a horizontal line, and then lay down the angles of elevation and depression, and the measured distances laid off along these lines: then the respective heights of the vanes on each staff being laid down on vertical lines passing through these points, will give the spots on which the staves stood, and thus the outline of the ground can be ascertained.

* *Manual of Surveying for India*, p. 549.

This method is chiefly used in running check levels which, as intimated above, are intended merely as a test of accuracy in more detailed work.

155. Difference of Heights by the Barometer.

We can of course obtain only approximate results by this method. Three causes tend to affect the work, (a) climate, (b) latitude, (c) season.

Thus in England the *diurnal range* of the barometer is scarcely perceptible owing to the wide fluctuations to which the instrument is subject, but in many parts of the world, e.g. in parts of the Mediterranean, the height of the barometer being known at 9 A.M. on two successive days, its height at any intermediate hour may be closely approximated to.

156. A detailed description of the Barometer has been already given in Chapter VI. (§§ 93—101). The following points need only be noticed in this place.

The height of the column is affected by two causes, (1) *Capillarity*, (2) *Capacity*.

The capillarity tends to depress the height*. In a tube of a diameter 0·1 inch the error due to this cause amounts to 0·07 inch, with a diameter 0·5 inch it amounts to 0·003 inch.

The error of capacity is thus caused;—if the mercury rises in the tube, the surface of mercury in the cistern must be depressed, and vice versâ. Hence it is necessary to determine in some way the zero point of the cistern surface.

The error is allowed for in *three ways* (§§ 93—96).

(1) By making the area of the cistern about 100 times that of the tube, and then the resulting error is so slight that it may be neglected.

(2) By shortening the inches of the scale in the required proportion.

---

* Vide Everett's *Text-Book of Physics*, p. 83. The top of the mercury ought always to be *convex*; if it should ever assume the *concave* form some imperfection exists in the mercury.

(3) By a flexible base, as in Fortin's barometer, the surface of the mercury can be always raised or depressed to the zero indicated by the ivory point.

It may be noticed also that a change of temperature of 1° F. causes an expansion or contraction of the mercury of $\frac{1}{9800}$ of its bulk. Hence in all delicate observations the temperature must be taken.

157. Method of using the Mountain Barometer.

In carrying it from place to place the most essential point to be borne in mind is that the instrument must be always kept in an *inverted position*. In reading we must make the index appear to touch the curved surface of the mercury. The height can be read off to the $\frac{1}{1000}$th of an inch (§ 99). The thermometer *attached* to the instrument, which shews the temperature of the mercury, and the *detached* thermometer, which shews that of the atmosphere, must always be noted.

158. The following is Dr Hutton's rule for computing the difference between the heights of two stations at which observations have been made.

(1) Correct the readings of the barometer, reducing them to the same temperature by increasing the colder, or diminishing the warmer, by $\frac{1}{9800}$ part for every degree of difference as shewn by the *attached* thermometers.

(2) Take the difference of the logs of the heights of the barometer thus corrected, and take the four figures of the difference to *the right of the decimal point* as integers.

This will be the approximate height *in fathoms*.

(3) Correct the number thus found for the temperature shewn by the *detached* thermometers as follows:

To every degree that the mean of the two differs from 31° take as many $\frac{1}{435}$ths of the fathoms found above, and *add* this quantity if the temperature be above 31°, but *subtract* it if the temperature be under 31°; the result is the *true altitude in fathoms*.

LEVELLING.

We will work the following example to illustrate the above method*.

	Corrected Bar.	Attached Therm.	Detached Therm.
At A	30·409 in.	61°	58°
At B	30·278 ,,	60°	57°

$$\frac{30\cdot 278}{9600} = 0\cdot 003$$
$$\underline{30\cdot 278}$$
$$30\cdot 281$$

Mean of Detached Therms.
$$= 57\cdot 5$$
$$\underline{31}$$
$$26\cdot 5$$

Log $30\cdot 409 = 1\cdot 483002$
Log $30\cdot 281 = 1\cdot 481170$
Diff. $= \overline{0\cdot 001832}$

∴ Cor. for Temp.
$$= \frac{26\cdot 5 \times 18\cdot 32}{435} = 1\cdot 116.$$

∴ Approx. Alt. in fms. = 18·32.

Cor. for Temp. = $\underline{1\cdot 116 +}$
$19\cdot 436$ or in feet = 116·7.

159. The following simple formula is found to give good results.

$$\text{Height in feet} = 55000 \times \frac{B - B'}{B + B'}.$$

Adding or subtracting $\frac{1}{140}$th of the result for every degree (Fahrenheit) that the mean temperature is above or below 55° F.; where B = Reading of Barometer at the Lower Station, and B' = Reading at the Upper.

The above example worked by this formula will give the height as 119 feet.

The following formula has been recommended in the case where the height does not exceed 2000 feet.

$$X = 52500 \left(1 + \frac{2\cdot \overline{T + T_1}}{1000}\right) \cdot \frac{B - B'}{B + B'},$$

where B = Height of the Barometer at the Lower Station,

* Frome, p. 116.

B' = Height of Barometer at the Upper Station,
T = Temperature (C.) at the Lower Station,
T_1 = Temperature (C.) at Upper Station,
X = Height in Feet.

The Aneroid is very useful in determining the altitude. Its results are considered good for at least 4000 feet; but it is necessary that this instrument should be very frequently compared with a Standard Mercurial Barometer.

160. DETERMINATION OF HEIGHTS BY THERMOMETER.

The Thermometer has been used to ascertain the height of a station approximately. It is well known that, as the pressure of the atmosphere diminishes, water boils at a lower temperature, and this fact has been pressed into service to discover the altitude of the place above the sea level. As a rule it is found that the heights so determined are somewhat *less* than those by the Barometer. The chief advantages of this method are, the great portability of the necessary apparatus, and its small liability to injury. The following appear to be the more important conclusions arrived at by Col. Sykes during his experiments in India:—When the Boiling Point is 212° (F.) the height is 0 feet; when 208° the height is 2050 feet; when 204° the height is 4130 feet; when 200°, the height is 6250 feet.

The Heights thus found are to be corrected by a multiplier, the value of which depends on the temperature of the air. When the temperature is 32° (F.) this multiplier is 1·000; when it is 45°, the multiplier is 1·027, and when it is 60°, the value of the multiplier is 1·058.

Note. If the height is calculated from the sea level it is said to be an *Absolute Height*, but if it depends on the assumed height of another station, it is known as a *Dependent Height*.

CONTOURS.

161. DEF. Contours are horizontal lines either round a group of isolated features of ground, or over an entire tract of country.

Let us suppose that a mountainous tract is covered with water, and that this water is gradually subsiding. Now if the water, at *vertical* intervals of 10 or 15 feet, could be supposed to leave a permanent trace on the sides of the mountain, these traces would give us a very clear idea of Contour Lines. They would (*a*) be perfectly horizontal, and (*b*) would trace out perfectly the sinuosities of the ground. Hence we should have marked out exact horizontal sections of the ground.

A good example of these Contour Lines is afforded by the Lines of Equal Soundings in a Chart.

162. The following is the method of tracing out Contour Lines round an isolated feature. The ground must first be very carefully surveyed, and those parts (ridges, water courses, &c.) which define the configuration of the surface must be marked out by pickets; the exact position of these pickets can be fixed on the plan.

A Spirit-Level is then placed so as to command the best view of the line of level, and a staff with a moveable vane is placed at one of the pickets; the vane is then raised or lowered until it is bisected by the cross wires of the Level. The staff, with the vane *kept at this height*, is then shifted to another point about the same level, and moved up or down the slope until the vane is again bisected, when another picket is driven in to mark the position. This process is continued until it becomes necessary to shift the position of the Instrument itself. Now manifestly the points where the staff rested are on the same level, and these points being accurately laid down in the Plan, the lines which join these positions will mark out a horizontal section of the ground, and be therefore Contour Lines. The same operation is necessary to form the Contours above and below that first laid out. When the vertical interval is small, the pickets which mark several Contours can be fixed without shifting the position of the Spirit Level. We require to have merely a levelling staff of sufficient length.

This method of delineating ground was introduced by French Engineers.

Suppose that *AB*, *CD*, *EF* represent three Contours, then we

11—2

infer that the ground is steep between A and E, but slopes away gradually in the direction BDF.

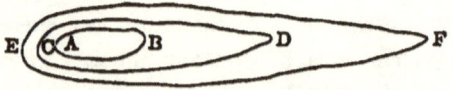

Shading is also used to shew the relative steepness of ground. The darker shading indicating regions where the ground is more abrupt, the lighter intimates slopes of a gentler description.

163. Mountain ranges are represented by fine lines, drawn nearly parallel at first, commencing at the summit of the hills, where they are usually drawn close together, and then they extend, as it were, to the level of the plains where they become more divergent. These lines are known as *hachure* lines, and are supposed to represent the courses which would be taken by rills of water trickling from the summit to the base.

EXAMINATION.

(1) Define *True Level, Apparent Level*.

(2) What do you understand by two points being on a level?

(3) Explain the difference between Simple and Compound Levelling.

(4) Mention the different methods of finding the vertical height of one point above another in levelling a district.

(5) Describe the process of conducting a Levelling operation between two distinct points.

(6) What is a Check Level? What is the use of such a Level? How is it usually obtained?

(7) Can you give any formula by means of which the difference of height between two points can be found approximately by the Barometer?

(8) What is a Section? Its use? How can it be formed?

(9) What are Contour Lines? Can you give an example from the Admiralty Charts?

(10) Explain the process of finding these lines.

(11) If a Chart or Plan were placed in your hand how would you know which hills were the steepest?

(12) Describe the use of the Level and Levelling Staves as applied to Nautical Surveying. (May, 1879.)

(13) An Aneroid barometer shewed at the sea-level 30·0 inches; and at the summit of a hill it shewed 29·35 inches; and finally on returning to the first station it marked 30·02 inches. If 85 feet of altitude $=\frac{1}{10}$ inch as shewn by the Aneroid, find the height of the hill. (Sept. 1879.) *Result* = 561 feet.

(14) Explain how the height of a hill may be ascertained by employing an Aneroid barometer. (April, 1880.)

(15) In the same level with, and equidistant from stations A and B, the height by Level was on the *back* staff at A 2·5 feet, and on the *forward* staff at B 7·2 feet. Shew by means of a diagram the difference of height of the ground at A and B.
(Aug. 1876.) B is 4·7 feet *below* A.

(16) At M equidistant from A and B, the height by Level on *back* staff at A was 3·2 feet, that on *forward* staff at B was 6·8 feet; shew by diagram the difference of the height of the ground at A and B. (Feb. 1877.) B is 3·6 ft. *below* A.

(17) Prove that the difference between the true and apparent level on the earth, regarded as a sphere, is always proportional to the square of the distance.

CHAPTER X.

TIDES AND TIDAL OBSERVATIONS.

164. BEFORE the time of Sir Isaac Newton, although many wonderful guesses were made as to the cause of the Tides, it may be very safely asserted that the subject was not in any wise understood. Thus we have the great Kepler writing, "The sphere of attractive virtue which is in the moon extends as far as the earth and entices up the waters, but as the moon flies rapidly across the zenith, and the waters cannot follow so quickly, a flow of the waters is occasioned in the torrid zone towards the westward*." So difficult indeed was the problem considered that it was proverbially known as "the grave of human curiosity."

Newton, however, applied his discovery of the Law of Gravitation with remarkable skill to the solution of the Tides difficulty, and explained satisfactorily the phenomena of the Neap and Spring Tides, and many other points connected with the subject.

Two principal theories have engaged the attention of mathematicians since Newton's time, viz. those of Bernoulli (about 1745) and Laplace (1774). Bernoulli imagines the earth to be a perfect sphere and covered throughout by water; then, he says, the waters would assume the same form, at any moment, which they would assume if the acting forces were invariable in magnitude and direction. In other words, taking account first of the moon's action, "a prolate spheroid of revolution in a state of

* *Elementary Treatise on Tides*, by Rev. J. Pearson, p. 4.

instantaneous equilibrium is imagined, its major and minor axes are assigned, and then solid geometry exhibits the mathematical relation between the length of a radius vector at any point, and the angles which fix its position, the constants being determined from observation. The like also is done when the sun is considered. The excesses above the mean radii in each case are superimposed, and thus a formula for the height is obtained. A similar investigation brings out the interval between the time of transit and that of high-water*." This is a statement of the famous *Equilibrium Theory*.

Laplace, however, takes a different course. He calculates the attractive forces of the sun and moon upon the ocean, and finds them to contain *constant* terms and *periodical* terms. He next states that in consequence of the resistance and friction of the waters they would soon have assumed a form of equilibrium under the forces which are represented by the constant terms, and then that the state of such a system of bodies is periodical when the forces themselves are periodical. In this way Laplace gets an expression for the height of the Tide the same as that obtained in the former theory†. This method is usually known as the *Dynamical Theory*.

165. Tidal Phenomena.

If any observer stations himself in any tidal harbour, he will perceive the following changes in the state of the water. He will observe that the water rises and falls twice each day; the rising of the water is the result of the *flowing* of the tide, the falling of the water is the result of the *ebbing* of the tide. When the water ceases to rise, the state of the tide is called *high water*, when it ceases to fall, it is called *low water*. The time of high water is about 40 minutes later on each succeeding day; and these times

* Pearson, p. 26.

† Pearson, p. 26. In his valuable work, Mr Pearson gives an interesting account of the labours of other eminent workers in this branch of Science. On page 27 will be found a description of the Harmonic Analysis of the Tides, a method introduced by Sir William Thomson. *Vide* also Thomson and Tait's *Nat. Phil.* Vol. 1. 479 for a description of a "Tide Predicting Machine."

of high water are observed to bear a close relation to the position of the moon, e.g. at Ipswich it is high water when the moon is nearly south; at London Bridge, when she is about south-west; and at Bristol when she bears about E.S.E. But these facts apply to only one of the high waters each day. The second high water occurs about $12^h\ 24^m$ after the first.

Again, the interval between the moon's passage over the meridian and the time of high water varies with the *moon's age*. At new moon, full moon, 1st quarter, and 3rd quarter (or rather the day which follows these four phases) the interval is nearly the same; from new moon to the 1st quarter, and from full moon to the 3rd quarter, the time of H. W. occurs *earlier* than would be inferred from using this same interval, whereas from the 1st quarter to full moon, and from 3rd quarter to new moon it occurs somewhat later.

During the days following new and full moon the Tide rises higher and falls lower than at any other period of the month, such tides are known as *Spring Tides;* but during the days following the 1st and 3rd quarters, the tide has the least *range* (*vide* definition given below § 183), and such tides are known as *Neap Tides*.

166. The tides are caused by the *inequalities of the attractions* of the sun and moon on the two sides of the earth, not to their *total* attractions. This fact must be very carefully borne in mind.

The accompanying diagram will shew how this differential action produces its effects.

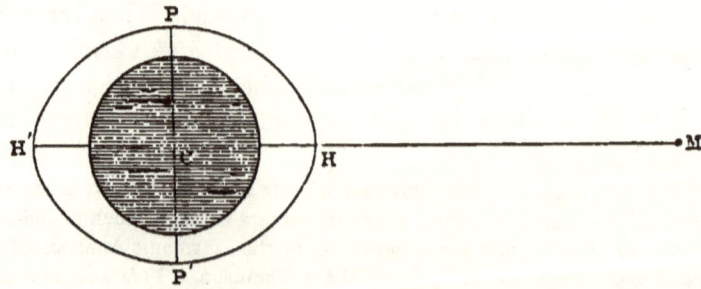

Let *M* represent the position of the moon. The shaded figure represents the earth, the centre of gravity being at C, and the plain envelope $HPH'P'$ the waters of the ocean, supposed to cover the earth completely.

Since the moon's influence diminishes inversely as the square of her distance*, it is plain that the waters at H are more attracted by the moon at M than the earth is at C, and again, the earth at C more than the waters at H'. Now as the earth must move as one mass, and the waters being fluid are free to obey the forces which act on them, it is evident that the waters at H will *rise* towards M, being drawn away, as it were, from the earth, and the waters at H' will also rise, being left behind, as it were, by the earth, which moves from them faster than they can follow.

167. Thus we can understand that there is high water *at the same time* on opposite sides of the earth. The tide of the hemisphere which has the moon above the horizon is called the *Superior Tide*, and the corresponding tide at the same time in the opposite hemisphere is called the *Inferior Tide*. Laplace supposed the obverse action of the moon to be produced by what he called an *Anti-moon*. Hence the Inferior Tide is called an *Anti-lunar*, or *Anti-solar* tide, according as it is produced by the obverse action of the moon or the sun †.

Again, as the total quantity of water is the same, it follows that a *rise* of the waters at H, H' must be accompanied by a *fall*

* i.e. at double the distance, her power is one-fourth, at three times the distance, her power is one-ninth, &c.

† If we take the sun's distance as 23142 times the earth's radius, and its mass as 314760 times that of the earth, the earth's action on a particle of water at its surface being represented by 1, then $\frac{314760}{23141^2}$ and $\frac{314760}{23143^2}$ will represent the sun's attraction on a particle on the sides of the earth adjacent to it and turned away from it respectively. In the case of the moon we have similarly $\frac{·0123}{59^2}$ and $\frac{·0123}{61^2}$. Hence the differential action in the case of the moon is much greater than in the case of the sun.

Lockyer's *Elementary Lessons in Astronomy*, p. 321.

According to Sir Isaac Newton the height of the Tidal wave due to the moon is to that due to the sun as 58 to 23, or about 5 to 2.

of the waters in other places; accordingly we find that there is a fall along the circle PP'.

168. In order that the water should attain the shape $HPH'P'$ in the above figure, the waters must flow away from P, P' towards H, H'. Hence if the whole earth were covered with a deep ocean, and the moon were placed at the point M, H being the point vertically under M, PP' being the great circle, whose plane is at right angles to the line MHH', we may say that there is high water at H, H' and low water at P, P'.

169. If the moon were fixed at M, the earth being at rest, and the influence of the sun neglected, the waters of the sea would eventually settle down into the above position of equilibrium, and all tidal motion would cease[*]. But, as a matter of fact, not one of these conditions is found to obtain in practice; the moon appears to move round the earth once in about $24^h\ 48^m$ (which period is called a *Lunar Day*), and hence the points of high water will travel with her, and the circle of low water will also move. The moon will pass the meridian twice *in a Lunar Day*, and we shall have two high waters and two low waters in every $24^h\ 48^m$.

170. The sun also produces an effect on the waters of the ocean, but though his mass is many times greater than that of the moon, his far greater distance is the cause of his inferior power in raising the waters (*vide* note on preceding page). When the two bodies are exerting their influence together, as at new moon, or in opposite directions, as at full moon, the tides produced by the coincidence of the tidal waves, are greater than usual, and form the Spring Tides; when the actions of these bodies tend to neutralize each other, as at the first and third quarters, the tides are lower than usual, and form the Neap Tides.

171. This, then, is a brief explanation of what the chief Tidal Phenomena would be on the twofold supposition: (1) that the

---

[*] The figure thus assumed is called the Tide Spheroid.

ocean covered the whole earth, and (2) that no friction existed between the waters and the bed of the ocean.

These conditions, however, as has been said, not existing, the above explanation will be of no use in estimating beforehand the probable course of events. In fact, so many and great are the modifications which exist, that the Tidal Phenomena are often extremely complicated, and, at the present date, notwithstanding the talent, skill, and labour that have been spent on the theory and observations of the tides, it is generally acknowledged that no other branch of physical science is in so unsatisfactory a state. We proceed, however, to lay down in as simple a manner as possible, such parts of the subject as are necessary for our purpose.

172. When the moon passes a meridian of a place, it is not high water *at* that place, but at some place about $30°$ behind the moon; in fact the time of high water follows the time of the moon's transit by an interval of time, greater or less, according to the retardation experienced by the water in passing from the open sea round islands, through narrow channels, and up rivers. This phenomenon of retardation of the times of high water is very well shewn in the accompanying "*Co-tidal*" map of the British seas on the days of new and full moon.

DEF. "*Co-tidal Lines*" are lines drawn through all the places at which it is high water at the same time. They thus represent *the form of the tide wave* which carries the crest of high water from one point to another.

173. The following explanation of the method by which the tides are produced will perhaps prove useful to the student.

Let M be the position of the moon, E the centre of the earth, P a particle of water. If PM represent in magnitude and direction the moon's attraction on P, and if X be taken in ME so that $\dfrac{MX}{MP} = \dfrac{MP^2}{ME^2}$, then XM will represent in magnitude and direction the moon's attraction on the earth at E.

Now conceive the earth reduced to rest with respect to the moon, by a force equal and opposite to XM applied to the earth, and to each particle of water on its surface, then the particle P will be acted on by a force PM, and by another equal and parallel

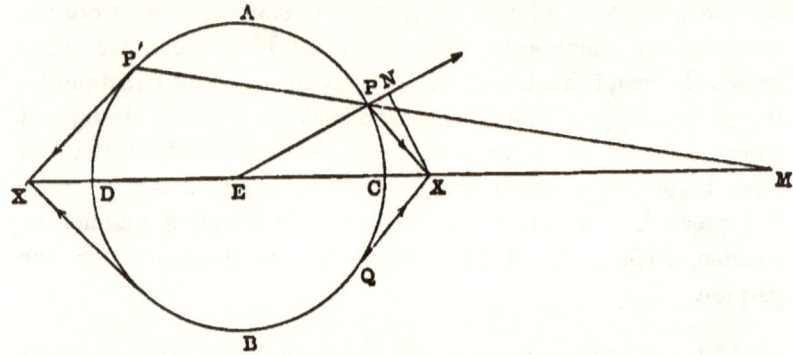

to MX, the resultant of which will be PX. Hence every particle of water in the hemisphere ACB will tend towards the line CM.

Again, if P' be a particle in the opposite hemisphere, and MP' represents the moon's attraction on it, and MX' be taken such that $\dfrac{MX'}{MP'} = \dfrac{MP'^2}{ME^2}$, then, by similar reasoning, the moon's disturbing action on every particle in that hemisphere ADB will be towards X' away from M.

Resolving the moon's disturbing effect into the normal and tangential components PN and PX respectively, their magnitudes may be realized as follows: Suppose the earth to be surrounded by water to the depth of 10000 fathoms, then the normal component PN will raise a tide of $\frac{7}{100}$ of an inch, and the tangential component PX will raise a tide of 58 inches. Hence it is easy to see what an important part the tangential force plays in the formation of the tides*.

174. Suppose that on the day of change, i.e., the day on which the moon is new, she crossed the meridian of Greenwich at noon, then, reckoning Greenwich time, the main features of the

* Brinkley's *Astronomy*, edited by Stubbs and Brünnow, p. 262.

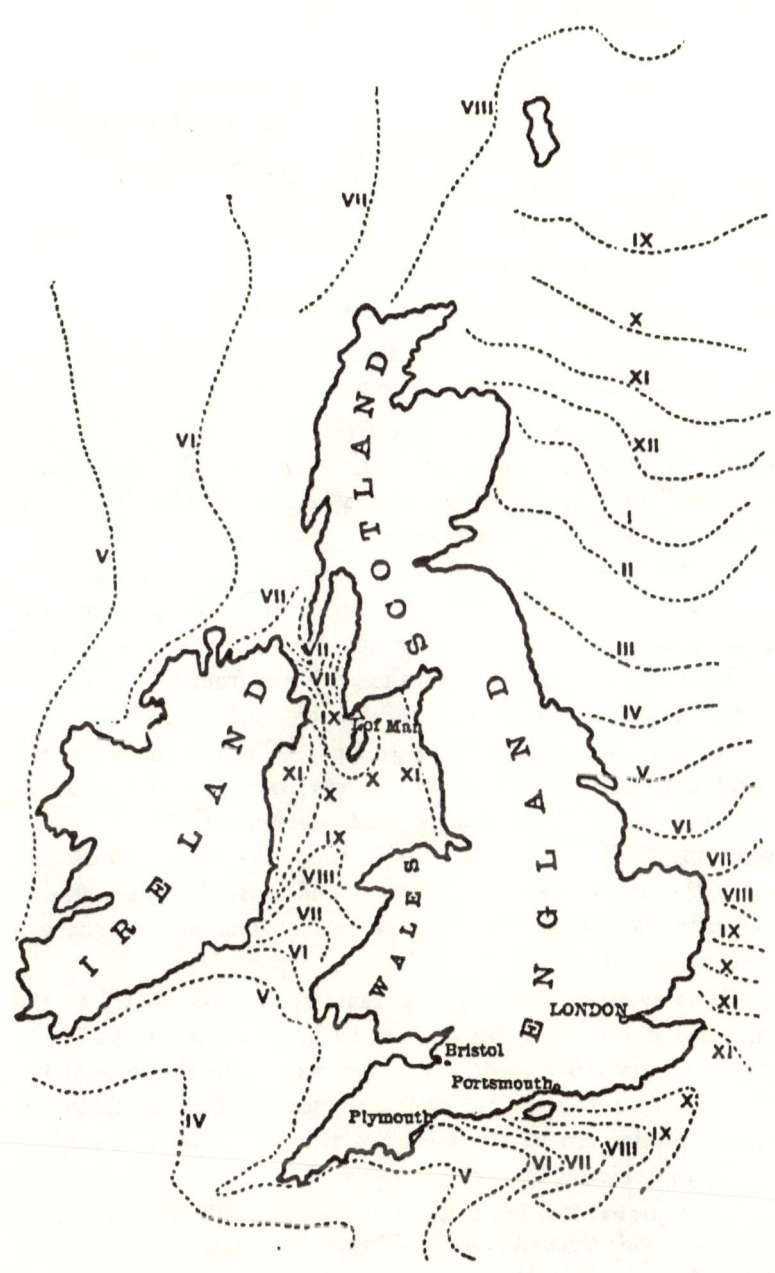

tidal wave (which has its origin in the Southern Ocean) coming up from the Atlantic are the following. The tidal wave brings H. W. to the western coasts of Spain and Portugal about 2 P.M.; to the western coast of France about 3 P.M. and at 4 P.M. we see in our map the 4 o'clock Co-tidal Line approaching the English and Irish coasts. At 5 o'clock it is H. W. on the west coast of Ireland, along the south coast of Ireland, at Scilly, along South Devon, at Ushant and at Brest.

At 6 o'clock the line of H. W. has been divided by Ireland.

(a) The northern branch causes H. W. along the shores of Mayo, West Donegal, and is approaching the shores of Scotland.

(b) The southern branch is divided by Cornwall; one part enters the Irish Sea, and makes H. W. at Wicklow in Ireland, and at Pembroke in Wales, the other part makes H. W. at Plymouth, Axmouth, and at Guernsey.

In this way we can trace the progress of the tidal wave. It reaches the straits of Dover 11 hours after the transit. The other branch which at 6 o'clock approaches Scotland, is divided by Scotland, one branch enters the Irish Sea from the north, and meets the branch which entered from the southward, between the Isle of Man and Co. Down in Ireland. The line in which the two branches of a tide meet in this way, is known as the *head or end of the tide*. We see that the branch which runs down the East coast of England meets, off the Thames, the branch which had come through the straits of Dover and derived from the same Atlantic wave about 12 hours previously, and here we have a second head or end of the tide*.

It is evident from the map that when it is H. W. at the entrance of the Irish Sea it must be L. W. between the Isle of Man and the Irish coast, and *vice versa*. Thus the level of the waters has, as it were, an oscillating motion, and the middle of this oscillation crosses the Irish Sea at a line called a *nodal line*. Along this there is little rise or range of tide.

* The greatest Rise and Fall of a Tide will generally take place along this line. *Vide* Galbraith and Haughton's *Manual of Tides and Tidal Currents*.

175. It is very important to distinguish between the apparent progress of the high water and the real progress of the water itself: in other words between the *Tidal Wave* and the *Tidal Current*.

The actual transference of the water itself is called the tidal *current* or *stream*, and is caused by the tidal wave being obstructed by shoals or banks. The *Tidal Wave* is not a circulating current, but is a broad and nearly flat undulation, the highest part being that which immediately follows in the track of the moon. On approaching shoal water this undulation is checked and a current is created, e.g. in the Irish Sea the current nowhere exceeds five knots an hour, in the Pentland Firth at spring tides it is as much as eight knots, and yet the tide wave itself as estimated by the co-tidal lines appears to move at a minimum rate of 20 knots and in some places on the coasts at 50 and even 100 knots. In the open waters of the Indian Ocean and the South Atlantic where its progress is not interfered with, the tidal wave moves with a very great velocity, in fact little short of that of the moon herself*. We infer therefore that the tidal wave may be compared to the undulations in a rope, or the shaking of a sail, or the fluttering of a flag. The movement only is transmitted. We can easily see this in the case of water itself by watching a floating cork. We see it at one moment on the vertex of a wave, at the next in a hollow; the wave has passed onwards, but the water immediately about the cork is left behind. Similarly with a boat in a heavy swell, the wave comes up threatening to hurry the boat before it, but merely passes under and onwards leaving the boat in much the same position with respect to a point on the bottom.

176. The *flow* and *ebb* of the tide are due to the alteration of the level of the water caused by the tide wave. When the water in a channel has been set in motion, in other words when a *tidal current* has been produced, the motion of the water does not

---

* Thus in the South Atlantic the velocity of the tidal wave is 700 miles an hour, and in Lat. 60° S., it is estimated at 670 miles.

immediately cease when it is either high or low water, but the momentum still continues to produce its effects. We must, therefore, remember that the direction of the current does not necessarily and in all cases change with the tide, but on the contrary will under certain circumstances continue to run for *some hours* after the time of high or low water. This is a very important point. Dr Whewell states* that great confusion has arisen from not distinguishing the *time of high water* and the *time of slack water ;* the latter means the time when the current changes its direction. Owing to this point not being attended to, many observations have been rendered doubtful and many valueless. Hence, then, it must be borne in mind that while the water runs in one direction it does not necessarily rise, nor while running in the opposite direction is it necessarily falling.

Sir George Airy in his *Tides and Waves* remarks that if an observer stations himself on London Bridge he will see that the water continues to run upwards even after the surface of the river has dropped two feet.

177. DEF. The period during which the tide is stationary is called the *stand of the tide*†.

DEF. If the *current of the flood* continues to run for three hours *after the time of high water*, the tide is called a "*tide and half-tide*"; if it runs for an hour and a half, it is called a "*tide and quarter-tide*"; and if it runs for three-quarters of an hour, it is known as a "*tide and half-quarter tide.*" This special feature of a "tide and half tide" is found in the Solent, and at some of the Channel Islands.

---

* In his Paper on "A First approximation to a Map of Cotidal Lines," contained in the *Phil. Trans.* for 1833, p. 157.

† For the sake of clearness, we will present in one view the definitions of these terms :

Slack water is the *cessation* of the current caused by the tide.
High water is when the level of the water ceases to *rise.*
Low water is when the level of the water ceases to *fall.*
Stand of the Tide is the *duration* of High or Low water without apparent change of level.

178. We shall here introduce some *definitions* which may be found useful before proceeding farther with the subject. The Moon is said to be in *conjunction* when she is on the same side of the Earth as the Sun, and when the centres of the three bodies are *in the same plane**; and she is said to be in *opposition* when she is on the opposite side of the earth to the sun, and the centres of the three bodies are in the same plane. When in conjunction, the moon is said to be *new*, when in opposition, she is said to be *full*. When the lines joining the centre of the earth with the centres of the moon and sun *are at right angles*, the moon is said to be in *quadrature*. This happens twice during a lunar month : when it takes place between new and full moon, she is in her *first quarter*, when it occurs between full and new moon, she is in her *third quarter*.

The following symbols are used to represent these positions of the moon :

    Conjunction ☌   Quadrature ☐
              Opposition ☍

When the moon is passing from conjunction to opposition she is said to *wax*, and when passing from opposition to conjunction she is said to *wane*. Finally she is said to be *crescent* in her first and fourth quarters, and to be *gibbous* in her second and third quarters.

The interval between two conjunctions, i.e. between two new moons, is called a *lunation*, and is exactly equal to $29^d\ 12^h\ 44^m\ 2^s \cdot 84$. The interval between conjunction and opposition, i.e. between new and full moon, is called a *semi-lunation*.

The paths of the heavenly bodies are not circles, but ellipses ; the sun being in one of the foci.

When the earth in its orbit is nearest to the sun (which takes place in January) it is said to be in *perihelion*, and when

---

* Though not necessarily in the same *line*. If the bodies are in the same line, we shall have an Eclipse—Total, Partial or Annular—of the Sun, at New Moon, and an Eclipse of the Moon—Total or Partial—at Full Moon.

it is farthest from the sun, which takes place in July, it is said to be in *aphelion*.

When the moon in her orbit is nearest to the earth, she is said to be in *perigee*, and when farthest from the earth she is said to be in *apogee*.

When the sun and moon are in conjunction or in opposition, their combined effect is greatest, and the tides thus produced are known as *Spring Tides*. When the two bodies are in quadrature the tides are least, and are known as *Neap Tides* (*vide* § 170).

179. *Note.* The spring tides do not occur *on* the days of new and full moon, but generally two, or three days later; similarly for the neap tides.

The spring tides are *highest* at the equinoxes (viz. in March and September), when the sun is close to the equator, and when the moon's declination is zero at full and change. They are *least* at the solstices (viz. in June and December), when the sun has greatest declination and the moon's declination is also a maximum and of opposite name to that of the sun. When the equinoctial spring tides occur, the combined influence of the sun and moon is exerted along the circumference of the greatest circle on the globe, while in the case of the solstitial tides their influence is exerted along parallels of smaller circumference. Other things being equal, it is evident that the tides of the *winter* solstice are higher than those of the *summer* solstice.

180. When the moon's declination is zero the tides are equally high in both parts of the lunar day. If the moon is not in the equinoctial, i.e. if her declination be either north or south, then there is a difference in the height of the A.M. and P.M. tides, and this difference is known as the *Diurnal Inequality*.

E.g. If the moon's declination is 20° N., then the summit of the superior tide is in 20° north latitude, and the summit of the inferior tide is in 20° south latitude, and in this case the tide in Lat. 10° or 17° will not be so high as when the moon's declination is 10° or 17°. The daily change in the moon's declination being considerable, a sensible inequality is thereby produced in the heights of the A.M. and P.M. tides.

The maximum of the Diurnal Inequality corresponds to the moon's greatest declination, although it may not appear until after the time of greatest declination. In like manner it disappears with the Moon's declination, but not until some time after she has crossed the equator; thus, at Liverpool the *age* of the Diurnal Inequality is six days, and at Singapore a day and a half.

Again, at Bristol, from the end of March to the end of September, the P.M. tides are 15 inches higher than the A.M. tides, but from Michaelmas to the end of March the A.M. tides are the higher.

"The tides at Kurrachee, Bombay, and probably other parts in India, are subject to a large Diurnal Inequality, which may accelerate or retard the times of H.W. sometimes to the amount of $1\frac{1}{2}$ or 2 hours, and increase or diminish the rise by a foot or more*." "The low water at Singapore is affected by a large Diurnal Inequality amounting at times to 6 feet†." "The tides on the coast of Tong King (S.E. China) are subject to a large Diurnal Inequality, one high and one low water generally occurring in the 24 hours‡." At King George's Sound in Australia there is a large Diurnal Inequality of the times which sometimes reduces the two daily tides to one§."

181. There is sometimes a large inequality in the times as well as in the heights, of the morning and afternoon tides. Thus near Cape Florida, this was found at its maximum in June, 1835, to amount to $2\frac{1}{2}$ hours‖.

Dr Whewell states that the Diurnal Inequality "affects in the largest degree the *time* of high water, and the *height* of low water¶." And Mr Parkes, who has investigated the tides of Bombay and Kurrachee, infers "that when there is no Diurnal Inequality in *high water time*, there is none in *low water height*,

---

* *Tide Tables*, p. 172 note. † Do. p. 175 note. ‡ Do. p. 177 note.
§ Do. p. 184 note. The Diurnal Inequality of the *time* means the difference between the *lunitidal intervals* of successive tides, *vide* § 191.
‖ Raper's *Navigation*, p. 322. ¶ *Phil. Trans.* 1840, p. 165.

and when there is none in *high water height*, there is none in *low water time*.*"

182. OTHER CAUSES WHICH AFFECT THE TIDES.

These are two in number, viz. The winds and the state of the barometer.

These affect both the times and the heights. Thus in the North Sea a strong N.N.W. gale and a low barometer raise the surface 2 or 3 feet higher, and cause the tide to flow all along the coast from Pentland Firth to London, half an hour longer than the times and heights predicted in the Tables. Again E.S.E. and S.W. winds produce opposite effects, which will be felt down the Channel as far as Dungeness; while, on the contrary, at the entrance of the Channel, at Plymouth, and as far as Portland, a S.W. wind with a low barometer will raise the surface, but a N.E. wind and a high barometer will always lower it†.

M. Daussy and Sir John Lubbock, on comparing the differences between the observed heights of the tides and their computed heights with the direction of the wind, concluded that the *effects of the wind* are insensible; but all practical men believe that these effects are considerable. In support of these latter, Sir George Airy states in his *Tides and Waves*, that on January 3rd, 1841, a gale "lowered the tides in the Thames 5 feet, and produced a depression of about 3 feet at Hull and at Dover, and a sensible effect at Bristol. At Dublin and at Glasgow the tides were raised by it‡."

M. Daussy (mentioned above), the eminent French hydrographer, was the first § to point out the influence of the atmospheric pressure on the height of the tide; he stated that a *low* barometer is accompanied by a *high* tide.

Sir John Lubbock remarks that a rise of 1 inch in the barometer is attended by a depression of 7 inches in the height of the tide at London, and of 11 inches at Liverpool. Mr Bunt,

* *Phil. Trans.* 1868, p. 688.
† *Tide Tables*, p. 113.
‡ *Ency. Metrop.* Vol. III. *Mixed Sciences*, p. 390.
§ In *Connaissance des Temps* for 1834.

who has paid much attention to the tides of Bristol, noticed a depression, under the same circumstances, of $13\frac{1}{2}$ inches in the tides of that port.

Finally, from an examination of the most carefully conducted tidal observations ever made, viz. those carried out under General Colby's directions during the Ordnance Survey of Ireland in 1842, the Astronomer Royal was led to infer "that a *negative* irregularity in the height of the barometer is accompanied by a *positive* irregularity in the height of the sea, 12 or 14 times as great as that of the barometer[*]."

183. THE RISE AND RANGE OF A TIDE.

DEF. THE MEAN LEVEL OF THE SEA is the middle between the levels of high water and low water at Springs[†].

Though the *heights* of H. W. and L. W. may vary to a considerable extent, yet the *mean level* does not vary much; e.g. in Singapore there is sometimes a difference of 6 feet in the heights of the tide, yet the mean level does not alter more than a few inches. The wind is the principal cause of this slight variation. The mean level may be found very closely by the observation of four *consecutive* tides which include the Diurnal Inequality.

DEF. THE RISE OF A TIDE is "*the vertical rise above the mean low water of Ordinary Spring Tides*[‡]".

Note. The *height* of a tide is the same as its rise, and the *datum line* is the level of low water of Ordinary Springs.

DEF. THE RANGE OF A TIDE is always *measured from low water of one tide to the high water of the following tide.*

Thus, in the case of a Spring Tide, its rise and range are identical[§], but not so in ordinary and Neap Tides. In all tides,

---

[*] "On the Laws of the Tides on the Coasts of Ireland," by Sir G. Airy. *Phil. Trans.* 1845.

[†] But see some remarks in Rev. J. Pearson's work, p. 31.

[‡] Ordinary, because at the Equinoxes the Spring Tides are higher than usual at high water, and lower than usual at low water.

[§] Restricting the name "Spring Tide" only to the highest tide in the semilunation, and the term "Neap Tide" only to the tide which has the least height.

except Spring Tides, the rise is always greater than the range; and this difference is a maximum at Neap Tides. The following diagram will, it is hoped, make these definitions clear to the reader.

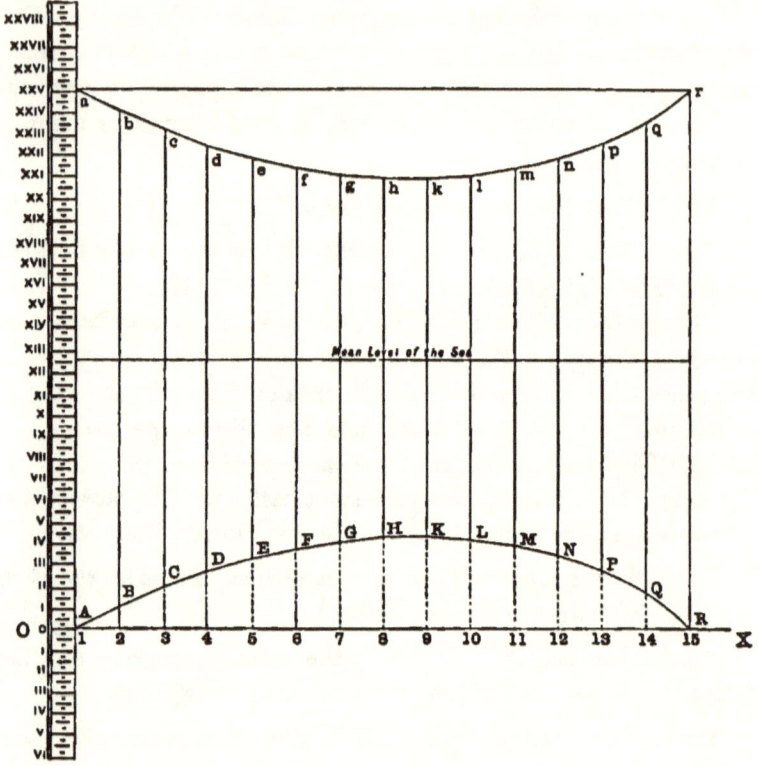

The graduated column to the left represents a tide gauge divided into quarter feet. The figures along the line OX represent the days of a semilunation, being the time from Spring Tides after new moon to the Spring Tides after full moon.

The lower curved line $ABC......R$ represents the curve of low water, while the upper curved line $abc......r$ represents the curve of high water.

Thus, let the line OX represent the line of low water at ordinary Spring Tides, and let the high water of ordinary Spring

Tides reach the height of 25 feet, marked a. Next day the water does not rise to that height, nor does it fall so low as A, let b mark the highest point reached, and B its lowest point; on the next day let c be the highest and C the lowest point of the tide; and so on, until the 8th day, when h marks the highest point and H its lowest. After this the highest points ascend and the lowest points descend continually until the 15th day, when we have again the phenomena of Spring Tides, and thus on continuously from Spring Tides to Spring Tides. On the 8th day of the semilunation we have the highest and lowest points of the Neap Tides.

184. The *Rise or Height of each Tide* is reckoned by its vertical rise above the line OX. Thus on the day of Spring Tide the rise is measured by the line $1a$, next day the rise is measured by $2b$, next day by $3c$, next day by $4d$, and so on, until, on the 8th day, the rise of the Neap is $8h$. After this we see the rise increasing day by day until we obtain on the 15th day the Spring Rise $15r$, corresponding to the height $1a$, 14 days before. We thus see that the Neap Rise is the least.

185. Again, the *Range of the Tide* on any day is estimated by the vertical distance between the two curved lines; e.g. on the 5th day the range is measured by the line Ee, on the 9th day by Kk, on the 13th day by Pp. It is also evident that the Neap range, on the 8th day, viz. Hh is the least. Hence the reader will observe why a distinction is drawn between the Neap rise and Neap range, but not between the Spring rise and Spring range. The diagram will also explain why the height and range both decrease from Springs to Neaps, and both increase from Neaps to Springs.

The *mean level of the sea* will be half-way between the curves of high and low water (§ 183).

186. The interval between the times of H.W. and L.W. is usually greater than the interval between the times of L.W. and H.W., and this is more marked at Springs than at Neaps. Thus at Liverpool the Flow occupies $5\frac{1}{4}^h$ and the Ebb $6\frac{3}{4}^h$.

DEF. A TIDE DAY is "*the interval between two successive arrivals at the same place of the same vertex of the tide wave.*" It varies in length as the two waves, due to the separate action of the Sun and Moon, approach to or recede from coincidence. In the first and third quarters of the moon the solar tide is to the westward of the lunar tide; in the second and fourth quarters it is to the eastward.

DEF. The lengthening and shortening of the Tide Day, thus caused, on its mean length is called the PRIMING and LAGGING OF THE TIDE.

The priming will evidently happen in the first and third quarters of the moon; the lagging in the second and fourth quarters.

187. DEF. SINGLE DAY TIDES are those which happen only once in 24 hours.

The most remarkable case is that of Tong King in China. Whewell thus describes it*: "The tide rises and falls every day during about 12 hours each way. The time of rise is three-quarters of an hour later each day, so that in 15 days the time of H.W. advances from 1 P.M. to midnight; after which it does not advance to 1 A.M., but falls back 13 hours to noon, and so on perpetually. In this way H.W. is always P.M. during the summer half of the year (March to October) and A.M. during the other half. When the tide time falls back 13 hours, the tides are scarcely perceptible; they are greatest at the intermediate times. Newton explained this by two opposing tides, one 6 hours longer than the other. When the Moon is on the Equator, the morning and evening tides of each component tide are equal, and the tides obliterate each other by interference, which takes place about the Equinoxes. At other periods the higher tides of each component daily pair are compounded into a tide which takes place at the intermediate times, i.e. once a day; and this will be after noon, or before, according to the time of the year."

* *Phil. Trans.* 1833, p. 224.

188. Def. Double Half Day Tides are tides which rise and fall *four times* in the 24 hours.

Poole in Dorset is an instance. Here, the tide ebbs and flows twice in 12 hours. It is L.W. at about $3^h 30^m$; then flows regularly until $5^h 20^m$, and makes proper H.W. about $8^h 50^m$; it then ebbs for $1\frac{1}{2}$ hours, and again flows for $1\frac{1}{2}$ hours, and finally ebbs until L.W.

This is a local circumstance, and is caused by the alteration of level by the velocity of the Ebb current near the shore; "and this alteration of level, from the hydrostatical effect of currents, shews itself in the form of a second rise of the surface, after it has begun to descend from the true H.W.*"

189. The highest Springs and the lowest Neaps in each lunation take place at different intervals after the moon's Syzygies† (i.e. conjunction and opposition) and Quadratures, at different places; the interval varying from a few hours to a few days. In like manner the tides *corresponding to* the Moon's *octants* (i.e. when her transits take place 3 hours after the Sun's meridian passage), follow the time of the octants at different intervals at different places. Such tides are called *Octant Tides* by Dr Whewell‡.

190. Establishment of the Port.

One or two preliminary Definitions are necessary.

Def. The Retard, or, **Age of the Tide,** is "*the interval between the transit of the Moon at which the tide originates and the appearance of the tide itself.*"

E.g. if the moon passes the meridian at 4 P.M. and the time of H.W. is 7 P.M., this tide does not, in general, *correspond* to the transit which took place 3 hours previously, but to a transit which may have happened a couple of days back: thus, on the west

---

* Whewell, Do. p. 226.
† This term is applied when the conjunction and opposition are spoken of together.
‡ *Phil. Trans.* for 1840, p. 260.

coasts of Spain and France the tide is 1½ days old, at London it is 2½ days old, on the west coast of Ireland 2 days, on the southwest coast of America 1 day 20 hours; whereas on the Pacific side of the United States it is scarcely half a day old.

The special transit to which any tide really corresponds is found by carefully examining the observations of several preceding tides; the highest of these, being due to the united influence of the Sun and Moon, must correspond to that transit of the Moon which took place at noon or midnight*. An extended series of very carefully conducted observations is, however, necessary to fix this with accuracy.

The term "Retard" explains itself: the tide appears to be "retarded" in following the Moon in her diurnal course. Bernoulli first applied the word in this connexion.

191. Def. The Lunitidal Interval is quite a different thing from the Age of the tide, and is defined as *"the interval which elapses between the Moon's transit each day and the time of H. W. next following."* It is a varying quantity from day to day during the semilunation. The irregularities are due to the angular distance of the Moon from the Sun, to the distances of Sun and Moon from the Earth, and to the changes in their declinations: the first is the chief cause.

Def. The Establishment of the Port is, it must be remembered, merely an *interval* of time, not, in the first instance, an hour of the day by any clock.

Def. The Vulgar Establishment of the Port is *"the interval between the time of the Moon's meridian passage on the day of new or full moon and the following high water,"* i.e. the *lunitidal interval* on the days of full and change †.

Def. The Corrected Establishment of the Port is *"the mean of the lunitidal intervals for all the days between new and full moon."*

* Raper, p. 320.
† The Vulgar Establishment is registered on the Admiralty Charts.

Note. If the new moon occurred exactly at noon or midnight of the place, then the Sun and Moon would pass the meridian at the same instant, and in this case the Establishment of the Port might be given as the time from apparent noon, but it is evident that this must be a very rare event.

When, therefore, it is said that the Change Tide or Establishment of the Port is $9^h\ 20^m$, it does not mean that on every day of new and full moon it is H.W. at $9^h\ 20^m$ P.M., but that it is H.W. $9^h\ 20^m$ *after the Moon's transit on those days*. In the event, however, of the moon passing the meridian at apparent noon, then, as has been explained, it will be H.W. at $9^h\ 20^m$ P.M. Keeping this in mind, we have given in the *Nautical Almanac* the following definition of the Establishment of the Port, viz. "The actual time of high water when the Moon passes the meridian at the same time as the Sun; or the interval between the time of transit of the Moon and the time of high water on full and change days."

DEF. The difference between the Corrected Establishment of the Port and the Lunitidal Interval at each transit of the Moon is termed the *Semimenstrual Inequality**.

192. The Corrected or Mean Establishment may be determined by observing the intervals between the times of the moon's meridian passage and the times of the following high waters for a semilunation, and taking the mean of them. To obtain the elements of a complete tide table, the high waters and low waters during an entire lunation ought to be observed. The most important lunations are those about the times of the Solstices and Equinoxes.

193. DETERMINATION BY MEANS OF CURVES.

To compare the time of H.W. with the time of the Moon's transit, we must compute the latter from the *Nautical Almanac*, and find how much the time of H.W. is after the Moon's transit at the

---

* Or, semi-monthly inequality. This term is due to Sir John Lubbock. Dr Whewell introduced the terms Vulgar and Mean or Corrected Establishments.

place. These differences are *lunitidal intervals*. Let us suppose that we have obtained the observations for H.W., we annex the times of the Moon's transit, and then put in the lunitidal intervals computed from these data. We thus obtain the following table*:

Day.	A.M. or P.M.	Times of High Water.	Times of Moon's Transit.	Lunitidal Intervals.
		h m	h m	h m
11	A.M.		10 33	2 34
	P.M.	1 7	10 57	2 32
12	A.M.	1 29	11 21	2 30
	P.M.	1 51	11 45	2 26
13	A.M.	2 11	0 9	2 20
	P.M.	2 29	0 32	2 16
14	A.M.	2 48	0 55	2 8
	P.M.	3 3	1 19	2 2
15	A.M.	3 21	1 42	1 54
	P.M.	3 36	2 6	1 48
16	A.M.	3 54	2 29	1 40
	P.M.	4 9	2 52	1 34
17	A.M.	4 26	3 15	1 28
	P.M.	4 43	3 39	1 24
18	A.M.	5 3	4 3	1 20
	P.M.	5 23	4 27	1 19
19	A.M.	5 46	4 51	1 18
	P.M.	6 9	5 16	1 18
20	A.M.	6 34	5 41	1 20
	P.M.	7 1		

We first put in the corrected time of the Moon's transit $10^h 33^m$. We next insert the time of the following high water $1^h 7^m$. Then take the first from the second (increased where necessary by 12^h), and the lunitidal interval $2^h 34^m$ is obtained.

Full Moon occurred between the 12th and 13th, because on the 13th the Moon was 12 hours from the Sun. Now by taking the times of H.W. as abscissæ and the lunitidal intervals as ordinates, we can ascertain whether a tolerably regular curve is formed, and therefore whether the lunitidal intervals follow a regular law.

Again, if we set off the times of the Moon's transit as abscissæ and the lunitidal intervals as ordinates we obtain a *curve of semi-*

* *Manual of Scientific Enquiry*, p. 66.

monthly inequality; and when this curve has been determined by observation for any place, the hour of H.W. at any time at that place may be predicted.

Once more, if we set off on this curve the ordinate which corresponds to the time when the Moon's transit is 0^h or 12^h, we obtain a graphic method of determining the Establishment of the Port.

Finally, if we set off the times of the Moon's transit as abscissæ, and the observed heights of the tides as ordinates, we obtain the *Law of the Heights of H.W.* from Springs to Neaps. The maximum ordinate (= spring height) follows the days of new and full moon by 1, 2, or 3 days, and, as the new or full moon is supposed to produce the spring tide, we thus get the *Age of the Tide*.

194. First and Second High Waters.

This is a purely local matter, and two very good examples are found in the Solent and at Havre. In the Solent "this double high water is probably caused by the tidal stream at Spithead, for, as long as that stream runs strong to the westward, the tide is kept up in Southampton water, and there is no fall of consequence until the stream begins to slack at Spithead; but when the stream makes to the eastward at Spithead the water falls rapidly at Southampton. After low water the tide rises pretty steadily there for 7 hours, which may be regarded as giving the first or proper high water; it then ebbs for about an hour, and falls 9 inches, when it again begins to rise, and in about $1\frac{1}{4}$ hours reaches its former level, and sometimes goes higher, this is called the second high water*."

At Havre, where the Spring Rise is 22 feet and Neap Rise is 18 feet, the high water remains stationary for about an hour, with a rise and fall of 3 or 4 inches for another hour; and during a total interval of 3 hours the tide only rises and falls 13 inches. This long period of slack water is very valuable for the traffic of

* *Tide Tables* for 1881, p. 110 note.

the port, and it allows fifteen or sixteen vessels to enter or leave the docks on the same tide*."

195. DEF. THE TIDAL CONSTANTS for any port are certain corrections of time and height, which being applied with their proper signs to the time and height of the tide at the *Standard Port*, will give the time and height of high water at the port required.

E.g. Portsmouth is the Standard Port of reference for all places on the south coast of England, from Littlehampton to Portland; Plymouth is the standard for all harbours from Bridport to the Scilly Isles; London for Gravesend, Woolwich, Greenwich, &c.

Examples. The Tidal Constants for Gravesend are:

$- 0^h 48^m$, and $- 3^{ft} 3^{in}$.

The Tidal Constants for Portland are $- 4^h 40^m$ and $- 6^{ft} 3^{in}$.

The Tidal Constants for Scarborough are $+ 0^h 49^m$ and $+ 1^{ft} 5^{in}$. Sunderland being the Standard Port for reference in this last case.

196. METHODS OF FINDING THE TIME OF H.W. ON ANY PARTICULAR DAY.

There are *three methods :*

I. By Stated Rules.

II. By Admiralty *Tide Tables.*

III. By the Tidal Constants.

197. I. BY RULE as given in treatises on Navigation.

(α) Find the mean time of the Moon's meridian passage on the given day.

(β) Correct this mean time by the equation of time to the nearest minute.

(γ) Take out approximately the Moon's semi-diameter.

* *Tide Tables* for 1881, p. 111 note.

(δ) Enter Table (l), page 5, in Inman's Tables, and take out the correction there found with its proper sign.

(ϵ) Apply this to the mean time of passage.

(ζ) To this add the Establishment of the Port.
Then—

i. If the sum $< 12^h$, the time is the mean time of the afternoon tide on the given day.

ii. If the sum $> 12^h\ 24^m$, or $> 24^h\ 48^m$, subtract these sums from it, and the remainder will be the time of H.W. on the afternoon on the given day.

(η) When the time of the P.M. tide has thus been found we can find the time of the A.M. tide on the same day by subtracting 24 minutes from it; and the time of the A.M. tide on the following day by adding 24 minutes to it.

This method gives only approximate results, and is now seldom employed.

198. II. By Admiralty Tide Tables.

This publication contains the Times and the Heights of the A.M. and P.M. tides at twenty-three standard ports in the British Isles, and at Brest, for every day of the year; and also the times and heights at full and change for the principal places of the globe. These Tables are computed in the Hydrographic Office, and give very interesting and valuable information about the Tides, especially those around our own coasts. In the *Tide Tables* for 1882 are given the tables of Lubbock, Whewell, and Pearson, by means of which the time and height of the tide may be computed.

The most important Tables are numbered from I.—X., and may be briefly noticed.

Table I. shews the semi-monthly inequality; or the interval between the Moon's transit, two days preceding a London Tide (and denoted by the letter B), and the time of high water; the Moon's parallax being 57′, her declination 15°; the Sun's parallax 8·8″ and declination 15°. For Portsmouth this constant is about $1^d\ 12^h$, for London $2^d\ 3^h$, for Hull $1^d\ 19^h$, for Leith $1^d\ 15^h$, &c.

Table II. gives the correction for the Moon's parallax. The arguments being the Moon's transit (B) and the H.P. The correction ranges between $+9^m$ and -9^m.

Table III. shews the correction for the Moon's declination. The arguments being the Moon's transit (B) and declination. The limits in this case are $+12^m$ and -12^m.

Table IV. shews the correction for the Sun's declination. The arguments being the Moon's transit (B) and Sun's declination. The limits are $+5^m$ and -5^m.

Table V. shews the correction for the Sun's parallax. The arguments being Moon's transit (B) and Sun's parallax, which ranges between $8\cdot94''$ and $8\cdot66''$.

The limits of the correction are $+3^m$ and -3^m.

These five Tables refer to the *time* of high water, and a second series of five similar Tables give the corrections to be applied to the *height*. In using some of these last Tables for other ports than London, the corrections must be multiplied by a certain constant which is proportionate to the difference between the semimonthly inequalities of the two places.

199. III. BY THE TIDAL CONSTANTS.

We merely apply to the times of the standard port the tidal constant for the time at the required port; the result is the time of H.W. at the port required.

E.g. Find the time of H.W. at Greenwich on December 14th, in the afternoon.

At London Bridge it is H.W at $7^h\ 37^m$ P.M.

Tidal Constant 0 15

∴ P.M. time of H.W. at Greenwich $= 7^h\ 22^m$

200. TIDAL OBSERVATIONS.

In every survey of a harbour tidal observations form a most important portion of the work. In the next chapter the subject of Soundings will be fully explained; we shall therefore here confine our attention to the methods of observing for the *Elements of a Tide Table*, which are :

(1) The Establishment of the Port.

(2) The rise and range of the tide throughout the semilunation.

(3) Any important circumstances which may thus become apparent in the rise and fall, the flow and ebb, of the tide, &c.

201. A TIDE GAUGE is a long batten, usually coloured black, with divisions painted in white, or cut into the wood. These divisions are feet and inches, or else feet and decimals of a foot. This is fixed firmly and vertically to a post driven into the ground, or fixed to a rock. The zero of the scale must be below the level of the lowest water observed in that spot in the harbour. If the harbour is long and narrow, several such gauges ought to be erected, and their positions carefully noted. A trustworthy observer must be stationed at each when the more important observations are being made. For very delicate observations in somewhat exposed situations, the following seems to be a satisfactory form of Tide Gauge. An upright tube, open at the top and closed at the bottom, with two or three small holes in the side near the closed end, is taken of a length somewhat greater than the rise of the Spring Tides at the place. This tube must be firmly fixed to an upright post driven into the ground or otherwise secured, so that the bottom of the tube may be always beneath the lowest water and its top above the highest water. The water is thus enabled to reach the same level inside and outside the tube, but the outside agitation of the waves does not sensibly affect the water inside. An upright rod attached to a float (which must be of nearly the same area as the section of the tube) will thus intimate very clearly the exact height of the water at any moment.

This may be slightly modified thus: a string from the float may pass over a pulley, and have a weight with an index mark secured to the other end. The index will mark on a graduated scale the rise and fall of the float, and therefore of the tide itself*.

202. Observations ought to be made every half hour at least, and every 5 minutes when the time of high and low water draws near. It is well to begin a series of tidal observations by noting the state of the tide every half or quarter of an hour during the whole day and night. In this way any unusual circumstance at once becomes apparent, such, for instance, as four tides in a lunar day, the rise and fall being checked or accelerated through any local causes, &c.

203. It is often very difficult to determine the exact time of either high or low water, not merely on account of the waves washing against the gauge (a source of error that can be eliminated in the ways described above), but by reason of the water stopping or hanging near the time of highest or lowest water, or else rising and falling irregularly. Thus Sir G. Airy, in discussing the tidal observations obtained by General Colby during the Irish Survey (§ 182), writes: "The difficulty of fixing on the precise time of high or low water will appear from this statement, that sometimes twenty or twenty-four successive observations (occupying a period of $1^h\ 40^m$ or 2^h) are registered with the same decimal of a foot for the height. The most perplexing case is that where the change of height, in respect to change of time, follows or may follow different laws before and after the principal phase. Thus at Limerick after L.W. the water sometimes rises as much in 10 minutes as it had previously dropped in 2 hours. It therefore appears right here, if several successive observations about L.W. are registered at the same decimal of a foot, to suppose that the real L.W. is a little before the last of those observations†."

---

* Sir G. Airy, in his Treatise already alluded to, when speaking with great approbation of the method pursued in tidal observations at Sheerness, thinks that if the waves outside are very boisterous, the waters might be admitted first into one side of a trough, by a few small holes near the bottom, and then through a second partition into the chamber where the float is placed. Thus all outside disturbance would be eliminated.

† *Phil. Trans.* 1845, p. 9.

204. The three following methods of procedure have at different times been recommended.

(1) Note the time when the water ceases to rise, and the time when it begins to fall. The mean of these two times will be the time of high water.

(2) A more satisfactory method is as follows:—At equal intervals of time (e.g. every 5 minutes) for about an hour near the time of H.W. observe the height of the water: either take the means of times and heights for the time and height of H.W., and similarly for L.W.; or, take the highest point reached for the height and the corresponding time for the moment of H.W.

(3) A third, and still better, method of dealing with observations made at equal intervals of time, may be adopted by taking the times of observation as abscissae and the heights observed at those times as ordinates, and thus form a curve.

Example. Find the Time and Height of high water from the following observations:

Times 0^h—0^m, 0—5, 0—10, 0—15, 0—20, 0—25, 0—30, 0—35, 0—40, 0—45, 0—50, 0—55, 0—60.

Heights 6^{ft}—0^{in}, 6—6, 6—6, 6—9, 6—10, 6—11, 7—0, 6—11, 6—11, 6—9, 6—5, 6—2, 5—10.

Now the selection of the greatest height 7^{ft}. 0^{in}. will give the time of H.W. as 0^h 30^m, but if transferred to paper*, and the line connecting the ordinates smoothed into a curve, the general run of the height will give the time of H.W. about 0^h 32^m, or 0^h 33^m.

This method of finding the time and height of the H.W. is called *Interpolating*.

---

* Messrs Eden, Fisher and Co., 50, Lombard Street, E.C., sell paper admirably adapted for curve tracing. The square inch is divided by faint blue lines into 64 parts. It is known as "Section-ruled paper," and may be had in different sizes.

The following is the form of the Tide Register supplied to Surveying Ships.

Register of Tides observed at .... in the Month of .... 188...

Day.	High Water.		Moon's Transit.	Bar at H.W.	Low Water.		Moon's Transit.	Bar at L.W.	Range of Tide.	Moon's Age.	Wind.	Turn of Stream.	
	Time H.M.	Height F.I.	H.M.	ins.	Time H.M.	Height F.I.	H.M.	ins.	F.I.	D.	Direct Force.	Flood H.M.	Ebb H.M.
A.M.													
P.M.													

(*Note.* The columns for the moon's transit are to be kept exclusively for the purpose intended, and are not to be otherwise occupied.)

Observers' Names.

205. It sometimes happens that there will not be any P.M. tide, because the lunar day is longer than the mean solar day. The mean interval between successive tides is $12^h 24^m 22^s$. Hence if the A.M. tide is at $11^h 50^m$, the next high water will not occur until past midnight, and this will be the A.M. tide of the following day.

Note. It may be well to call attention here to the Four very remarkable Papers on Tidal Evolution, by Professor G. H. Darwin, published in the *Philosophical Transactions* for the years 1879 and 1880. The Mathematical investigations are of the very highest order; but the ordinary reader will find the "Review of the Tidal Theory of Evolution as applied to the Earth and the other members of the Solar System" (pp. 879—885, Part II. 1880), in which the Professor gathers up his conclusions, sufficiently popular to give him a deep interest in the important part played by the Tides in the past history of our Earth. The student ought also to read the *Phil. Trans.* for 1879, Part II. pp. 532—537, and especially Dr Ball's pamphlet "A Glance through the Corridors of Time" published by Messrs Macmillan and Co.

EXAMINATION.

(1) Explain, by means of a diagram, the phenomena of two simultaneous tides in opposite hemispheres.

(2) Explain the causes of Spring and Neap tides.

(3) What is an Octant tide?

(4) Under what conditions are the spring tides greatest?

(5) Distinguish between a tidal wave and a tidal current.

(6) Explain the following terms:
> Rise and Fall of the tide.
> Flow and Ebb of the tide.

(7) Explain the following terms:
> Slack water.
> Stand of the tide.
> Head of the tide.

(8) Define the following terms:
> First of the flood.
> Last of the flood.
> First and Second high waters.

(9) Define the following terms:
> Tide and half tide.
> Tide and quarter tide.
> Tide day.
> Priming and lagging of the tide.

(10) Define very carefully the following important terms:
> Change tide.
> Vulgar Establishment of the Port.
> Corrected Establishment of the Port.
> Lunitidal interval.
> Semimenstrual inequality.

(11) Define, and illustrate by a diagram, the following tidal terms:
> Rise of a tide.
> Range of a tide.
> Height of a tide.
> Rise and range of a neap tide.

(12) What are the "elements" of a tide table?

(13) What is the diurnal inequality? What causes it?

(14) Define the "mean level of the sea."

(15) What is a lunation? Give its length.

(16) What term is applied to the interval between new and full moon?

(17) When is the moon said to be new, and when is she full?

(18) Explain the following terms: Quadrature, Perihelion, Aphelion, Perigee, Apogee, Solstice, Equinox.

(19) Write down the symbols for Conjunction, Opposition, and Quadrature.

(20) What do you understand by the Superior and Inferior tides?

(21) Explain the meaning of the Tidal Constants of a port.

(22) Describe a tide gauge, and the precautions which ought to be observed in its erection.

(23) If the observations must be made where waves are likely to prove a source of uncertainty, what method is to be pursued?

(24) Explain very fully the method of making tidal observations, and note the seasons at which these observations are most important.

(25) I find this entry in the official Tide Tables, "Half of the mean Spring Range is 9ft. 4in." Explain this.

(26) If, on a certain date, I find in the Tide Tables that the time of H.W. at a place is $6^h 5^m$; is this mean or apparent time? Is it (for England) local or Greenwich time?

(27) The height of the tide is 11ft. 6in.; what is the Datum line?

(28) What are Single Day tides?

(29) Explain Double Half Day tides.

(30) Explain this sentence: "The tides of Avatcha Bay are affected by diurnal inequality."

(31) Explain this sentence: "The times of H.W. being the corrected and not the vulgar establishment."

(32) Explain the abbreviation "H.W. F. & C."

(33) How is the time of H.W. calculated for any particular day when the Change Tide is known?
(Exam. Papers, March, 1880.)

(34) Does the diurnal inequality affect the time or the height of the tide?

(35) But other causes may require a different answer. Shew that this is so.

(36) Does the direction of the tidal current always change at the time of H.W.? (Exam. Papers, April, 1880.)

(37) When do the highest tides occur? (Sept. 1880.)

(38) State how the time of H.W. by a tide gauge is accurately determined, and in what time (apparent or mean) it is expressed. (Dec. 1875.)

(39) Explain how the mean tide level is ascertained.
(Sept. 1876.)

(40) If the tide continues to flow for $\frac{3}{4}$-hour after the time of H.W., what term is given to the tide?

(41) How would you find the time and height of H.W. at a home port by means of the Tidal Constants?

(42) State the effect of a rising and falling barometer respectively on the tides.

CHAPTER XI.

SOUNDINGS.

206. ONE of the most important parts of the survey of any harbour consists in obtaining accurate soundings in its every part. Two objects must be kept in view during the operation, viz. the *depth of water*, and the *nature of the bottom*.

207. In sounding it is advisable to take casts in straight lines. In shoal water a *sounding pole* may be used. This instrument is a long pole graduated to feet and inches, or feet and decimals of a foot. A flat piece of wood at the bottom prevents it sinking into the ground. In deeper water the lead line, carefully graduated, must be employed. The boat must be kept on a straight line by the method described in the next paragraph, and in this way the presence of shoals, rocks, &c., can be detected.

208. The boat must be fixed at starting by two angles, in fact by the "three-point problem." Note the direction in which it is intended to sound, and if two natural objects in line in that direction are not to be had, then recourse must be had to staves with flags (§ 50), the position of these artificial objects being, however, accurately fixed. And similar marks must be erected along the shore on the lines in which the soundings are to be made. Keeping these marks in transit, the boat is pulled at a uniform speed, and at regular intervals casts are made. If the survey is to be a close one, a cast ought to be made each half minute in water under five fathoms in depth, and each minute in water of a greater depth*. The boat's position ought to be fixed at

* Some surveyors prefer to space the soundings by the number of strokes of the oars rather than by time intervals.

given short intervals of time and always fixed at the end of each line of sounding. The boat is then pulled parallel to the shore until the next line is reached, when she sounds in a direction parallel to the former line. In this way the soundings can be fairly *spaced* on a chart, provided that the speed of the boat is uniform and the casts are made at equal intervals.

Note. A line must be drawn under all soundings which were made when the tide was up, and which when "reduced" (vide below) to low water would be apparently on a *dry* bank (p. 16).

209. In soundings the "*quality of the bottom*" ought to be always examined, and its nature accurately described on the chart by the side of the figures denoting the depth (§ 1). In foggy weather the nature of the bottom is often as much a guide as the depth. A remarkable illustration occurs in nearing the English Channel on the parallel of 49° 20′ N. The same depth of 73 fathoms occurs at spots 125 miles apart, but in the *outer* depth the bottom is "fine sand," while in the *inner* depth it is "ooze," and this distinction in foggy weather is the seaman's only guide*.

210. In keeping the boat in a line by means of "two objects on with each other," the two objects ought to be separated by a considerable interval, as, if too close, the plan is not sufficiently *sensitive*, and the boat might deviate considerably from the required line before the error would be noticed. It is also an important thing to remember that the position of these objects must be accurately protracted on the chart.

211. To FIX A SOUNDING, we must use either the "three point problem", or the "straight line and one angle method." These two methods have been so fully described in Chapter IV. (§ 50 ff.) that nothing further need be said here.

If a shoal or other danger is discovered, it ought to be at once fixed by the "three-point problem," great care being taken in the selection of the three "points."

212. During the progress of the sounding operations, the lead lines ought to be frequently tested in their *wet state*, as two

* *General Instructions for Hydrographic Surveyors*, 1877, p. 8 note.

causes tend to change their length, viz. (1) *constant wetting*, and (2) the *strain of the lead*. The watches supplied to the boats, and to the observers at the tide gauge, ought also to be compared every morning and evening: this precaution will ensure accuracy in the after process of reducing the soundings to the required datum line.

213. REDUCTION OF SOUNDINGS. All soundings taken during a survey, must be reduced to the *level of low water at ordinary spring tides* before they are entered on the chart; the object of these soundings on the chart being to let the seaman know the *least* depth of water he can reckon on at any state of the tide. This reduction is thus effected :—

While the soundings are being taken in a boat, an observer is ordered to note carefully the depth of water at the tide gauge, at intervals of 5 or 10 minutes, and to register his observations thus * :—

Time.	Depth of Water at the Tide Gauge.
H. M.	FT. IN.
10 0	17 6
10 10	17 9
10 20	18 0
10 30	18 4
&c.	&c.

The depth of water at each sounding is noted by the officer in the boat, and also the *time* of each cast. This information is thus registered:

Time.	Depth of Sounding.
H. M.	FT. IN.
10 0	63 6
10 10	73 0
10 20	59 8
10 30	43 3
&c.	&c.

* These examples are, of course, only imaginary, and intended merely to explain the method of proceeding.

Now, let us suppose that the lowest water observed at any period during the survey, by means of the tide gauge, was 5 ft. 8 in., then we conclude that the plane of the *low water level of ordinary springs* intersects the tide gauge at 5 ft. 8 in. above the zero of the scale. Hence, at 10 o'clock, the level of the water in the harbour was 17 ft. 6 in.—5 ft. 8 in., or 11 ft. 10 in. above the *datum line* to which the soundings are to be reduced.

At 10 o'clock a sounding of 63 ft. 6 in. was struck. Therefore, to "reduce" this sounding to the given datum line, it is evident that we must subtract 11 ft. 10 in. We thus obtain 51 ft. 8 in. as the depth of water to be marked on the chart. The other soundings are reduced in the same way.

It is usual to make a further reduction of a couple of feet in cases of uncertainty, and where such reduction can be effected without inconveniencing the traffic of a harbour or port.

214. Perhaps a diagram will render clearer the method of reducing soundings.

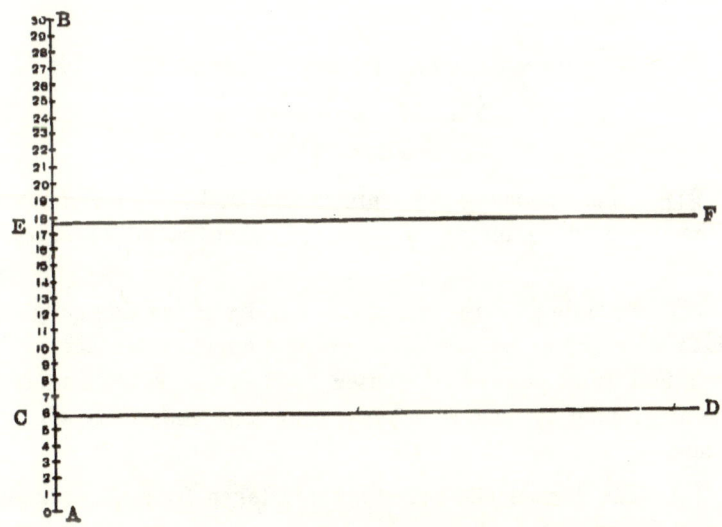

Let AB represent the tide gauge marked to 30 feet.

Let CD represent the level of low water at ordinary springs intersecting the tide gauge at 5 ft. 8 inches. Hence, we infer that under ordinary circumstances, CD will represent the plane of the lowest water in the harbour.

Let EF mark the height of the water, suppose 17 ft. 6 in., when a certain cast has been made; then the plane of water in the harbour at that instant was higher than the datum line by the vertical distance EC. Now $EC = 11$ ft. 10 in., therefore 11 ft. 10 inches must be subtracted from the sounding made at 10 o'clock, whatever it was, in order to reduce it to the datum line.

215. Formula for reducing Soundings.

Let s = depth of a sounding at a certain hour.

g = depth of water at the tide gauge at the same time.

d = depth of water at the tide gauge at low water of ordinary springs.

x = depth of sounding to be registered on the chart*.

Then $x = s - (g - d)$,

$\quad = s - g + d$,

$\quad = s + d - g$.

Thus, to reduce the second sounding taken at 10·10 o'clock,

$$x = 73\text{ ft. }0\text{ in.} + 5\text{ ft. }8\text{ in.} - 17\text{ ft. }9\text{ in.} = 78\text{ ft. }8\text{ in.} - 17\text{ ft. }9\text{ in.} = 60\text{ ft. }11\text{ in.}$$

$\therefore x = 60$ ft. 11 in.

216. The following rules are recommended by a competent authority as desirable to be observed when sounding in tidal rivers†.

(1) Soundings, in the immediate vicinity of the tide gauge by which they are to be corrected, are not appreciably affected by deviations from parallelism caused by the river water, and such soundings may be taken at any time and under any circumstances.

(2) The farther the soundings are taken from the standard gauge the greater is the probable error arising from non-parallelism.

---

* Subject to any further reduction that may be deemed advisable.

† Stevenson's *Canal and River Engineering*.

(3) The soundings ought to be made in *neap* rather than in *spring* tides*.

(4) The soundings ought to be made in *ebb* rather than in *flood* tides.

(5) Soundings taken in a flood tide, especially during springs, should not be made till within an hour or so of the time of high water.

Note. The parallelism of a river is affected by two causes; the rise of the tide may be great, or, the current of the river may be very strong.

217. Mr Mitchell Henry, of the United States Coast Survey, has proposed the following method of determining the elevations along the course of a tidal river without the aid of a level.

Set up graduated rods at such distances apart that the "slacks" of the tides may extend from one to another. By simultaneous observations ascertain the difference in the readings of these gauges at the "slack" between the ebb and flood, and again between flood and ebb, then apply the formula:—

Difference of elevation of the zeros = one-half the sum of the differences of the readings at the two slack waters.

218. To fix on a chart the position of a hidden danger which is at a considerable distance from the shore, we must anchor a boat or a buoy over it, and then fix the position of the boat by angles taken to objects on the shore.

If the danger is so far off that it can be viewed from only two points on shore, then its exact position may be laid down as follows. Anchor a boat or a buoy between the source of danger and the shore, and then the "danger" can be fixed by the "three-point problem".

219. When an eddy, or still water, intimates danger, the following method of procedure may be adopted if the state of the sea will permit. A buoy is anchored where the danger is

---

* Some surveyors adopt the very opposite rule to this.

supposed to exist and the buoy's position is fixed by the "three-point problem." Soundings are then taken round the buoy in ever-enlarging circles until the source of danger is discovered.

The boat can also be anchored at the head of the eddy and then dropped slowly astern by the boat rope until the point of danger must have been passed. She is then hauled up and again dropped astern on another line, and so on, until the danger is found. When this occurs, the boat anchors to *windward*, and sufficiently near the danger that on dropping astern by her boat rope she may reach the position necessary to be fixed.

Or, finally, the danger may be found by the method of "Sweeping." Two boats pull abreast at a certain distance apart. Hanging from each boat's stern is a heavy weight, and a deep-sea lead line connects these at any required depth. Thus it is scarcely possible for any known danger to escape detection.

220. A Surface Current.

The existence of such a current becomes manifest almost at once. Its *rate* and *direction* may be thus determined. Allow a barrico to drop astern by the log line, or else the log ship itself: this determines the *rate*. Observe the bearing of the barrico by the standard compass, and thus the direction can be determined.

The rate of the current can also be determined by means of the *Current Log*. In this instrument the logship is considerably larger than in the ordinary log used at sea. The line is marked at intervals of 10 feet, and is allowed to run out for a certain number of *minutes*. The following formula will give a very approximate result.

Let v = rate of the current in knots per hour,

f = number of feet run out,

t = time in *minutes*,

and let the length of the knot be 6000 feet.

Then $\dfrac{v \times 6000}{f} = \dfrac{60}{m}$,

$$\therefore v = \dfrac{60f}{6000m} = \dfrac{f}{100m}.$$

Hence the practical rule adopted is "Divide the number of feet run out by 100 times the number of minutes; the quotient will be the rate of the current in knots per hour."

e.g. In 4 minutes 360 feet ran out,

$$\therefore v = \frac{360}{400} = \cdot 9 \text{ knot.}$$

221. Under-currents.

The *existence* of such currents, as well as their *rates* and *directions*, are usually determined in the following way. An "Undercurrent Float" is made either of two pieces of sheet iron in the form of the letter T, and suspended by cords so that it may remain horizontal, or consists merely of a weighted basket with pieces of sail-cloth fixed on it so as to catch the water*. This float is then suspended by a line of sufficient length to allow it to descend to any required depth, and is suspended from a light block-tin buoy of such a shape as to offer least resistance to the water. This float serves the double purpose, (1) of keeping the underfloat at the same depth, and (2) of indicating the course and strength of the under-current. Smaller floats are thrown out to gauge the upper current, if any, and to ascertain the distance traversed by the buoy in a certain time.

Note. The under-current ought to be allowed to get up its proper rate in the float before the smaller marks are let go. Under the most favourable conditions we can only hope for approximate results.

Captain Spratt, in his paper before the Royal Society in 1871, gives an account of observations on the under-currents in the Sea of Marmora; he found that the best float for the purpose was one made of thin copper or block-tin, suspended so as to remain horizontal, and a buoy anchored with a sinker, served to shew the relative speed of the surface and under-currents.

---

* The first kind was used by Stevenson in 1875 in the survey of Cromarty Firth, the second kind by Carpenter and Jeffrey in 1870, during the cruise of H.M.S. "Porcupine."

222. Mr Mitchell Henry, already alluded to, has used successfully the following method of discovering the existence of an under-current. A tin cylinder about 40 feet long and 3 inches in diameter is made in separate sections which can be rendered perfectly air-tight. Each section, however, is provided with a stop-cock, so that the instrument may be filled to sink it to any required portion of its length. As the tube drifts nearly upright in the water, with its top protruding a few inches above the surface, its velocity may be taken as the *mean motion* of the stream; if it leans backwards or forwards it shews that its foot rests on a stratum of water which has greater or less motion than the surface drift; and, finally, if its course differs from that of a surface float, the action of an under-current is recognised, the direction and force of which may be approximately determined.

223. THE DISCHARGE OF RIVERS.

DEF. A stream is said to be in its *normal* condition during *ordinary* summer weather, i.e. when it is neither dried to its minimum by a long drought, nor swollen to its maximum by heavy rains.

DEF. By the *discharge of a river* is meant the quantity of water which passes out of the river in a given time; it is generally expressed in the number of cubic feet per minute.

This may be computed from the formula:

Discharge = *sectional area* × *mean velocity*.

We have, therefore, to devise methods for ascertaining the sectional area, and also the mean velocity.

224. TO FIND THE SECTIONAL AREA.

Select a part of the river where the banks are regular and the stream tranquil, and stretch a graduated cord across the stream. The depth of water can then be found with a sounding pole, or lead line, at every 5 or 10 feet along the cord; thus the *mean depth* can be ascertained. Then we have

Sectional area = *mean depth* × *width of the river*.

DEF. The average depth of water at a section is known as the *Hydraulic Mean Depth*.

225. To find the Mean Velocity.

The most accurate method seems to be to ascertain the surface velocity in the middle of each of the compartments into which the transverse section of the river is divided by the soundings, made as already explained, and then the *mean velocity* of these may be taken as representing the mean velocity of the river.

Two methods are adopted for finding the surface velocity: (1) *by Floats*, (2) by an instrument called a *Tachometer**.

(1) *By means of Floats.* A float is let go at such a distance that when it reaches the line of section the stream is exerting its full influence on it. Its time of passage between two points, at a known distance apart, is then noted, and hence the velocity of the stream may be approximately computed.

(2) *By the Tachometer.* The principle of this instrument is much the same as that of the Patent Log. The instrument is fixed at any point, and the current impinging on a vane causes it to revolve. The number of revolutions made by the vane being registered on an index, the velocity of the stream is indicated. The instrument is applied at each compartment of the cross section, and the mean result taken. The great object in view is to ascertain the velocity of the current as it passes the line of section fixed upon.

226. The Ground Log consists of a logship with a sufficient amount of stray line to permit the logship being anchored at any point. A hand lead line, divided like an ordinary log line, is bent to the logship. The logship remains stationary in the water, and hence the direction and speed of the boat over the bottom, i.e. the course and distance made good, can be found.

Note. In greater depths than 4 fathoms it is advisable to use the "Buoy and Nipper." In this instrument the stray line runs through a notch until the lead reaches the bottom, when it is caught, and the Buoy remains stationary: then, as before, the speed of the boat can be found.

* *i.e.* a speed-measurer.

The Ground Log is chiefly used in river surveying, and in shoal water, especially where the vessel may be out of sight of land, or where the shore presents no distinct objects by which the position may be fixed.

Examination.

(1) In making a sounding what two objects must be kept in view?

(2) In sounding over a bay what precautions ought to be taken to insure the detection of all dangers?

(3) What is meant by "spacing" soundings on a chart?

(4) How may a boat be kept on a "range" while sounding?

(5) How would you fix a sounding on a range?

(6) In the case of a very important sounding having been struck, how would you proceed to fix it?

(7) What two causes tend to make the lead lines erroneous?

(8) To what datum line are soundings reduced in the Admiralty charts?

(9) Describe fully the method by which this reduction is effected.

(10) Investigate a formula by means of which this reduction may be computed.

(11) Suppose a tidal river has to be surveyed, what instructions as regards the soundings would you frame for the guidance of the observers?

(12) What do you understand by the parallelism of a river?

(13) How would you fix the position of a rock visible from only two points on shore?

(14) The reported position of a sunken rock in a surveyed harbour is given; how would you proceed to verify its existence, to fix it accurately, and to find the least water on it, the rise of the tide being considerable? (March, 1875.)

(15) Explain the principle of the "Ground Log," and state the description of navigation in which it is mostly used.
(March, 1878.)

(16) Give any simple methods you are acquainted with for ascertaining the rate and direction of a surface current from a ship at anchor. (March, 1878.)

(17) A sunken pinnacle rock, of small area, having 13 feet of water upon it, is reported in a vicinity in which the current runs 5 knots. Wishing to fix its position accurately, how would you proceed to find the rock? (May, 1878.)

(18) State how you would find approximately the quantity of water passing down a river, through any particular section of the river, in a given time. (Dec. 1879.)

(19) What methods are adopted to ascertain from a ship at anchor the velocity and direction of surface and under currents?
(Oct. 1880.)

(20) How is the sectional area of a river found?

(21) What is meant by the "hydraulic mean depth"?

(22) In what units is the discharge of a river expressed?

CHAPTER XII.

CHRONOMETERS.

227. DEFINITIONS :—

A *time-piece* simply marks the time.

A *clock* shews the time and strikes the hour.

A *watch* is a pocket time-piece.

A *repeater* is a clock or watch which by mechanism can be made to repeat the hour.

A *chronometer* is merely a very perfect watch, or time-piece, in the construction of which no skill is spared; and its mechanism is such that any change of temperature will produce the least possible effect on its performance.

228. *History of the Chronometer.*

It seems that striking clocks were known in Italy as early as the 13th century, or the beginning of the 14th century. In England, in 1288, a fine imposed on the Chief Justice of the King's Bench, was devoted to the purchase of a clock for the famous clock-house near Westminster Hall. In 1523, the church of St Mary in Oxford was provided with a clock from the proceeds of fines imposed on the undergraduates. About 1360, Edward III. gave protection to three Dutch clock-makers who were invited by him from Delft. We have some reasons for thinking that clocks were becoming well known in England at the end of the 15th century. Much discussion has taken place about the inventor.

It is now generally acknowledged that it was not the product of any single mind, but the result of many inventions. Thus, in the case of our modern chronometers, successive improvements and inventions have brought them to their present state of perfection.

In 1484, we hear first of a balance-clock being used for astronomical purposes: the principle of this instrument was that a finely-adjusted beam balance moving from side to side caught successively the teeth of a wheel and thus regulated the motion *.

Such seems to have been the success attending these experiments that we find an astronomer named Frisius proposing, about 1530, the use of a portable balance-clock for ascertaining the longitude.

In 1560, Tycho Brahé, the famous Swedish astronomer, possessed four clocks which shewed the hours, minutes, and seconds. The largest had only three wheels, the diameter of one of which was three feet, and had 1200 teeth on its rim. Tycho seems to have been the first to notice the effects of temperature, but apparently did not know how to explain the facts.

In 1577, an astronomer named Moestlin had a clock which beat 2528 times in an hour, and by its means determined the Sun's semi-diameter on passing the meridian: he made it to be 34′ 13″.

Clocks were certainly reduced to a portable size previous to 1544, and before this could have taken place the spring must have superseded the heavy weights. This may be considered as the second stage in the development, and prepared the way for the *Fusee*. Galileo, in watching a chandelier hanging by a long chain in a church at Florence, noticed the isochronism of the pendulum. The invention of the pendulum clock marks the third stage in the history. As usual in cases of this kind the inventor is doubtful. Huyghens applied the invention in a most skilful manner, and hence has been looked upon as the originator. It is now, however, known that in 1641, a London maker, Richard Harris, invented and constructed a pendulum clock.

* Vide description and diagram in Godfray's *Astronomy*, p. 36.

Soon after this invention was made known, Huyghens attempted to construct a marine clock, but his success was not great. He noticed that the pendulum beats slower as the latitude is diminished, and thus prepared the way for the correct knowledge of the Figure of the Earth.

In 1680, a London watchmaker, named Clement, invented the "anchor" escapement*. This change led to the mode of suspending the pendulum by a thin and flexible spring. The seconds pendulum, with this escapement, was known as the Royal Pendulum.

In 1715, George Graham, another London maker of great repute, endeavoured to obviate the effects of temperature by means of his well-known *Mercurial Pendulum*†. The famous John Harrison, by his *Gridiron Pendulum*, improved on Graham's invention. Graham then introduced his "dead-beat" escapement as an improvement on the anchor or recoil escapement. From the days of Harrison and Graham successive improvements have been introduced into every part of the mechanism.

229. Harrison's Watches.

During the reign of Queen Anne, in 1714, the British Government offered a reward of £20,000 for *any method* by which the longitude could at all times be determined at sea; the whole reward would be given if the method, when tested by a voyage to the West Indies, were found true within 30 miles, £15,000 if true within 40 miles, and £10,000 if true within 60 miles.

Harrison came to London, in 1728, with drawings of a watch which he deemed would answer the purpose. Dr Halley, the well-known English astronomer, to whom he applied, referred him to Graham, who soon discovered his great ability, and advised him to actually construct his machine before making application to the Board of Longitude.

In 1735, he presented his first watch, and next year was sent to Lisbon to test its power in a voyage. He corrected the dead

---

* Vide description and diagram in Godfray's *Astronomy*, p. 41, and also in Ninth Edition of *Encyc. Brit.* Vol. VI. p. 17.

† Vide Godfray's *Astronomy*, p. 42.

reckoning about *a degree and a half**, a success which naturally obtained for him both public and private help.

In 1739, he finished his second watch, which, though not tried at sea, gained him still further encouragement. In 1749, he presented his third watch †; it was less complicated and more accurate than the second, as its error was only 3 or 4 seconds a week. Thus encouraged he constructed his *fourth* and most famous chronometer, and applied for the full reward. The tests were very severe. The chronometer was compared in the Observatory at Greenwich, sealed up, and sent to Portsmouth, where Robertson, the Master of the Royal Academy, having found its error by equal altitudes, forwarded the observations and results to the Admiralty. The chronometer was then put on board H.M.S. "Deptford," commanded by Captain Digges. It was secured by four separate locks, the keys of which were entrusted to Governor Lyttleton, who was proceeding to Jamaica, to Captain Digges, to the Senior Lieutenant, and to Harrison's son. The ship sailed on Nov. 18th, 1761. During the voyage to Madeira the chronometer corrected the dead reckoning, which was sometimes in error to the extent of a degree and a half ‡. The ship arrived at Madeira three days before H.M.S. "Beaver," which had sailed ten days before her, a result, according to Harrison's account published in 1767, "which was owing to the Beaver being deceived in her reckoning by trusting to the log for want of a more perfect method of finding her longitude."

In going from Madeira to Jamaica, the chronometer corrected the errors of the longitude, which occasionally amounted to *three degrees*, and the reckoning of several ships in the convoy varied

---

* The Master of H.M.S. "Orford" in the homeward voyage thought that the point of land sighted was the Start, but Harrison, trusting to his timepiece, insisted that it was the Lizard, and he was found to be right.

† This gained the gold medal of the Royal Society.

‡ "In sailing to the Madeiras, Mr Harrison acquainted Capt. Digges with the time when he would see the Island of Porto Santo; which had they trusted to the ship's reckoning, they could not have seen in that voyage, which would have been a great inconvenience to them, *as they were in want of beer*"! Vide *Account of the goings of Mr J. Harrison's Watch*, 1767.

five degrees from the correct position! When the ship arrived at Port Royal, the longitude as found by the chronometer was only five seconds of time in error. The same precautions were taken during the return voyage to Europe, and on arrival at Portsmouth the error in longitude was less than 18 miles.

In the spring of 1764 the same chronometer was sent on a voyage to Barbados. Before sailing, Harrison drew up a declaration as to the rate by which he was content to abide. If the temperature was 42°, he said his machine would gain 3 seconds in 24 hours, if 52° the gain would be 2 seconds, if 62° the gain would be 1 second, if 72° there would be neither gain nor loss, and if 82° it would lose one second in 24 hours. In 156 days it was actually found to have gained only 54 seconds, allowing the chronometer to have gained 1 second a day, being the rate by which Harrison would abide; but if allowance be made for the change of temperature according to the above scale, the chronometer lost only 15 seconds! a result which shews a "wonderful performance."

Harrison, after some trouble and anxiety, obtained the whole £20,000, and a further sum from the East India Company.

230. The Inside of a Chronometer.

A chronometer derives its power from a coiled spring; the variable force of which is rendered uniform by the *fusee*, a beautiful contrivance, by means of which, on the principle of a variable lever, the main-spring acts through the medium of a chain. On the fusee is cut a curve, into which the chain fits, and which has this peculiar property, that as the chain winds upon it, the distance from the centre of motion of the fusee to the axis of the chain, at the point where it leaves the fusee for the barrel, continually varies; but this variation is such that the product of this distance and the force of the main-spring acting along the chain at that instant is constant, i.e. shall be the same wherever the chain leaves the fusee. Hence, then, we see that the power of the fusee to turn the machinery is always the same, and since the main wheel which communicates motion to all the rest is attached to the fusee, their centres of motion being coincident, it follows that the power at the teeth of the main wheel is uniform. This power is transmitted through the medium of a train of wheels and pinions till it comes to the Escapement.

Thus far there is not much difference between the works of a chronometer and an ordinary pocket watch.

231. The distinguishing features of a chronometer are the Escapement and the mode of *Compensating* the Balance for temperature. The latter being the more important for our purposes we shall devote some space to it.

AA is the balance arm. BB the two segments of the rims attached at one end to the arm and having the other end free. Each segment of the rim is composed of two metals, *steel* on the *inside*,

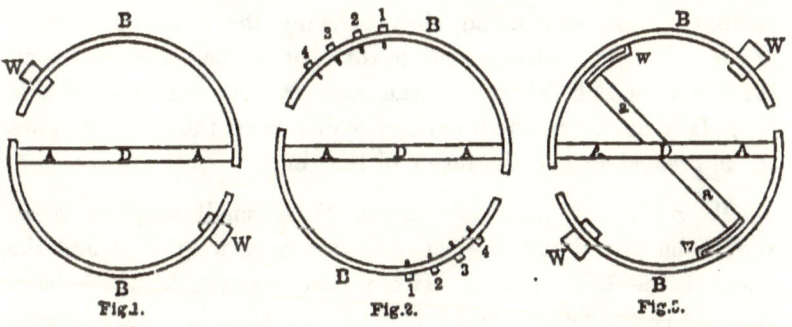

Fig.1.   Fig.2.   Fig.3.

and *brass* on the *outside*. W, W, are two weights which can be fixed at any points on the rims. Any increase of temperature *diminishes* the elastic force of the balance-spring, and hence the chronometer would lose, but, since brass expands more than steel, a curvature of the rims takes place inwards, and therefore the weights W, W, approach the centre. The inertia of the balance is thus lessened, and the balance-spring exerts the same influence as it did before the change of temperature. Again, if the temperature falls, the elastic force of the balance increases, and hence the chronometer would gain, but, since brass contracts more than steel, the rims curve outwards, the weights W, W, are therefore removed to a greater distance from the centre. The inertia of the balance is increased, and the balance-spring has no greater influence than it had before. It is evident that the nearer the weights W, W, are to the free ends of the rims, the greater will be the space through which they move by any change of temperature, and therefore the greater the variation in the inertia of the balance; hence it follows that if any change of temperature causes a change of rate the compensation is not sufficiently active. In Fig. 2, the small screws answer the same purpose as the weights in Figs. 1 and 3. In Fig. 3, we have represented the method of making the final adjustment of the "Thermal Compensation," invented in 1875 by the late Astronomer Royal, Sir George Airy, and which is now widely adopted by the best makers. It may be thus described:—

AA is the balance arm turning on the staff D. BB the ordinary brass and metal rims carrying the ordinary weights W, W. The new attachment consists of a small arm aa which turns with a stiff friction on the staff D. At each end of this secondary arm, are small springs which keep the little weights w, w, pressed against the inside of the rims.

To make the final adjustment, these small weights, w, w, should be placed midway between the large weights W and the ends of the balance arm. If the compensation by these large weights is insufficient the small weights w must be brought nearer to W, and vice versa. In this way no other adjustment is disturbed.

232. However perfect may be the compensation of a chronometer for certain temperatures, it does not hold good for all temperatures. If compensated for great heat and great cold, it *gains* at medium temperatures; on the contrary, if specially compensated for the latter, then the chronometer is out in either extreme. Chronometer makers at once sought for a remedy, and this remedy, as in Airy's method, is called the "secondary compensation," and the error, which it is intended to obviate, is known as the "secondary error." In the Royal Navy, the chronometers are usually compensated for 65° F., with a range of about 15° on either side of the mean.

To encourage the manufacture of good chronometers, the Admiralty, in the years 1822—1835, presented prizes for the excellence of individual chronometers. Since 1835 the prizes have been withdrawn, but the "annual trials" still continue. In successive years different makers obtained the post of honor, Poole, Frodsham, Hutton, Hewett, Eiffe, Dent. During the five years 1845—1849, no less than 219 chronometers were exhibited, and 79 were bought for the public service; the highest price given being £62.

233. CHRONOMETER ROOM IN THE ROYAL OBSERVATORY.

This is situated in the same building as the Great Equatorial, and is octagonal in form. On the left side of the entrance, are arranged the government chronometers and on the right the "annual trials*." These chronometers are so arranged on double rows of shelves placed back to back, that the superintendent and his assistants can make the daily comparisons at 1·30 P.M. without moving the instruments. There are usually about 250 chronometers in the room.

A galvano-magnetic clock by Shepherd is fixed to the central pillar. When the mean solar clock of the observatory is corrected each day for the 10 o'clock A.M. signal sent to the General Post Office, this clock is also compared, and after 1 o'clock P.M. the chronometers are compared.

It was formerly the custom to expose the "annual trials" to both a *cold* and a *heat* test. The chronometers were exposed in the open air on the coldest days in winter, but now, only a few of the very best are sometimes placed in the outside air, for a short time, to make further trial of their performances. The *heat test*, however, still continues.

Occupying one side of the room is the apparatus for this purpose. A zinc or galvanized iron case contains within it

* No maker can send more than two chronometers to compete in these annual trials. Each maker's instruments are kept together.

13 gas jets, the noxious products of the combustion being carried off by a pipe which communicates with the open air. A wooden case with glass and wood covers is arranged over this stove, and is fitted with shelves for the reception of about 80 instruments. The wooden cover being raised, the chronometers can be compared through the glass, and thus the temperature is not disturbed.

A thermometer shews the degree of heat, and a vessel of water being also placed inside, the cases of the instruments do not suffer from the excessive dryness of the air. The temperature ranges from 95° to 105°.

The "annual trials" remain in the Royal Observatory for six months, viz. from January until July. During this period they are subjected twice, for four weeks on each occasion, to the heat test, generally in March and May; and, as a rule, it is expected that with a good instrument the "weekly rates" (= sum of the "daily rates") when in the stove and out of it ought not to differ by more than 10 seconds. Formerly 20 seconds were allowed, but with the recent improvements for compensation the limit has been reduced.

When the chronometers are placed in the heat case a Chronometrical Thermometer (technically known as a "Chron. Therm.") is also enclosed. This instrument differs from an ordinary chronometer in having the two metals of the compensating rims reversed, *brass* being the *inside* metal, and steel the outside. The result of this arrangement is that the effect of any change of temperature is enormously increased, e.g. when the temperature in the chronometer room was 37°·4 the Chron. Therm. gained in a week 2194^s·8, when the temperature was 45° it gained 1624^s·5, and when in the heat case the temperature was 95°·9 it *lost* 3056^s·4.

234. In the published rates of chronometers, on trial for purchase by the Board of Admiralty, at the Royal Observatory, the results arrived at are arranged in three tables.

(1) The weekly sums of the daily rates in the *order of time*.

Note. If the weekly sum exceeds 40 seconds, the compensation of the instrument is looked to by the Officials in charge of the chronometers.

(2) The weekly sums of the daily rates in the *order of temperature*, as determined by the Chron. Therm.

(3) The abstract of the principal changes of rates.

The chronometers are arranged at the close of the six months in the order of their "Trial Numbers." These trial numbers are computed from the formula

T. N. = *twice the greatest difference between any two successive weeks + difference of the greatest and least weekly sum*.

E.g. In the trials for 1880, Cornell's chronometer had as the greatest weekly difference $3^s \cdot 7$, and the difference between the greatest and least weekly sum was $16 \cdot 3$.

$$\therefore \text{T. N.} = 2 \times 3\cdot 7 + 16\cdot 3 = 23\cdot 7.$$

The next in order was Kullberg's. Greatest difference $4\cdot 4$; difference between greatest and least $14\cdot 9$.

$$\therefore \text{T. N.} = 2 \times 4\cdot 4 + 14\cdot 9 = 23\cdot 7.$$

The third was Matheson's. Greatest difference $5\cdot 6$, and difference between greatest and least $13\cdot 9$.

$$\therefore \text{T. N.} = 2 \times 5\cdot 6 + 13\cdot 9 = 25\cdot 1.$$

The worst was No. 16494. Greatest difference $= 24\cdot 1$, and difference between greatest and least $= 57\cdot 2$.

$$\therefore \text{T. N.} = 2 \times 24\cdot 1 + 57\cdot 2 = 105\cdot 4.$$

The first three or four chronometers on the list have a price put on them by the Officials at the Royal Observatory somewhat in excess of their market value, and these are purchased by the Admiralty. The average cost is about £37 or £38. Sometimes an instrument is bought for £33, but these low-priced chronometers are only purchased when they are really wanted.

The chronometers in the observatory are compared every day at 1·30 P.M. A skilful observer will compare 200 chronometers in 16 minutes. Two copies of the rates of the government instruments are kept, one for the information of the Hydrographer, the other for use in the Royal Observatory. When the service chronometers are found to be going satisfactorily, they are marked "Ready" in the register, and the Hydrographer selects from his copy those instruments which are to be forwarded to the dockyard depôts, or when a ship is commissioned. The "Numbers" of the chronometers thus sent out from the Royal Observatory and their Makers are kept in this form:

Date.	Name of Maker and Number.	Ship.	Officer's Name.	When Delivered.	Remarks.

E.g. H.M.S. Northampton, which commissioned in 1880, had delivered to her on October 2nd, the following chronometers: Molyneux 2119, Mitchell 621, Dent 2209, Cairns 792, Eiffe 3, Pennington (D. W.) 1312.

The account kept at the Royal Observatory of the repairs of a government chronometer is thus registered:

Name of Maker and No. of Chron.	Whether two or eight Days.	Date of Last Repairs.	Maker to whom Notice of Repair is sent.	Date of Notice to Maker.	Day of Removal.	Notes on the Examination.	Date of Estimate.	Nature of Repairs.	Amount of Estimate.	Date of Authorizing Estimate.	Date of Report to the Admiralty.	Date of Return of the Chronometer.

235. HISTORY OF A CHRONOMETER WHILE IN USE IN THE ROYAL NAVY.

From what has gone before, the reader will know by what means a chronometer is admitted into the government service, and how it is selected for use afloat by the Hydrographic department of the Admiralty. The extracts, given below, from the

Observatory records will shew the "life" a chronometer leads. The first chronometer was returned to the Observatory from H.M.S. Formidable in 1866.

When a chronometer has been in use for three years, and is on the home station, it is generally sent to be examined by the maker. When its performances are no longer trustworthy, it ceases to be retained in the service, and is usually given to the maker, with a sum of money, in exchange for a new instrument. The old chronometer is often transformed into a Deck Watch. The cost of keeping a chronometer in repair ought not to exceed by much one pound a year. A chronometer, for instance, which was in active service from 1852 until 1875, cost the government £29 for repairs.

A new Balance Spring costs about ten pounds.

To whom Transferred, Maker, Ship, or Depot.	Date of Transfer.	From whom Received.	Date of Receipt.
Dent	Sept. 10, 1866	Dent	Oct. 15, 1866
Serapis	July 1, 1867	Portsmouth	April 13, 1870
Dent	April 18, 1870	Dent	Oct. 31, 1870
Topaze	June 24, 1871	Devonport	Nov. 1, 1872
Dent	Dec. 9, 1872	Dent	Sept. 22, 1873
Tamar	June 17, 1874	Devonport	May 13, 1875
Dent	May 24, 1875	Dent	Jan. 31, 1876
Hotspur	June 9, 1876		

Example II.

		Sheerness	Sept. 26, 1866
Dent	Oct. 1, 1866	Dent	May 20, 1867
Dent	Feb. 3, 1868		

Note. Given in February, 1868, to Dent with £26 in exchange for Dent 2945.

236. When a chronometer is being transported the following precautions are taken*. The brass case containing the chronometer is removed from the jimbals, the glass case is unscrewed, and the instrument taken out. The balance is then secured by two thin wedges of cork, and the chronometer replaced, the glass case screwed on, and the whole wrapped up in a sheet of thin paper. The jimbal ring is next removed, and the screws placed in a small packet in the bottom of the case. Some soft stuffing (such as dry oakum, paper shavings, hair, &c.) quite free from dust (sawdust or wood shavings are *not* to be used) is then laid at the bottom, and the ring placed on it. Over the ring place more stuffing, then the chronometer is laid in carefully, and secured with more stuffing up to the glass lid of the wood box. The case must then be enclosed in a wicker basket, or, if necessary, in a large quantity of stuffing sewed up in stout canvas, so that under no circumstances can the instrument receive a *jarring* blow. This precaution is of the greatest importance.

237. METHOD OF TRANSPORTING A GOING CHRONOMETER FROM THE SHORE TO A SHIP.

The instruments are secured from swinging in their jimbals by clamps provided for the purpose, and are generally carried between two men in a sheet of canvas. The object of carrying them thus is twofold, (1) to prevent their receiving a *jarring* blow, and (2) to keep them free from any *circular motion*. Villarceau recommends the following method: a trustworthy man stands within a hoop about a yard in diameter; he then takes up two chronometers properly secured, one with each hand, and thus safely carries them, free from any circular motion, and also secured from knocking against his person when walking.

238. STOWAGE ON BOARD.

When placed on board the following conditions are, as far as possible, sought for in stowing the chronometers.

(1) Where the traffic is least. (2) Where the tremor from the machinery is least felt. (3) Low down in the ship where the

* *General Instructions for Hydrographic Surveyors*, p. 34.

motion is least. (4) Where damp cannot attack them. (5) Where the ordinary temperature is most uniform. (6) Not near large masses of iron on account of *possible* magnetic influence*.

The following *negative* precautions ought also to be observed:

(1) The instruments ought never to be removed from their position except when the ship is in dry dock, or for extensive repairs; nor (2) ought they to be placed on swinging tables, or on a bed during firing, or held in the hand, or placed in drawers.

The errors arising from these various supposed precautions are, in fact, greater than the errors from those causes which they are supposed to guard against.

239. In the management of chronometers, the two things to be chiefly guarded against are *damp* and *variable temperature*. The former is a prolific source of injury to the instruments; e.g. here is a remark taken from the "Register of Repairs of Government Chronometers" in the Royal Observatory, when the chronometers are returned from ships paying off, "Chronometer dirty. Pendulum spring rusty."

240. EFFECT OF TEMPERATURE.

The general tendency of change of temperature may be thus

---

* Villarceau remarks that ships now-a-days are composed almost entirely of iron, and that experience seems to prove that the magnetism of the ship has little influence on the rates, on the supposition that the parts of the chronometer itself are not magnetic.

If a chronometer does present this phenomenon it is quickly detected by the variations produced by the changes in the direction of the ship's head. On the other hand, Chauvenet states that the "rates of chronometers have been found affected by masses of iron in their vicinity, thus indicating a magnetic polarity of their balances. Such polarity may exist in the balance when first it comes from the maker, or it may be acquired by the chronometer standing a long time in the same position with respect to the magnetic meridian. In order to avoid any error that might result from this polarity (whether known or unknown) it will be well to keep the chronometers always in the same position."

The reader may also consult Shadwell on Chronometers, pp. 14, 15 and notes, for further examples of divergence in opinions on this subject.

stated:—An *increase* of temperature causes the chronometer to *lose*, and a *decrease* of temperature causes it to *gain*. (§ 231.)

Besides the temperature, the performance of a chronometer depends generally upon the *age of the oils* used in the works. Of course only the very best oil is used, but after a certain time it thickens, and the effect of this is to diminish the amplitude of the vibration of the balance, and thus the chronometer is accelerated; and this acceleration is found to be almost exactly proportional to the time since it was freshly oiled. There is danger, however, in applying fresh oil, as, when the chronometer is subjected to the heat test, the oil often runs away from the pivot where it is required, and spreads over the plate where it thickens.

When the chronometers are received on board, they are arranged in a situation combining as many as possible of the above favourable conditions. The dial plates are placed in one direction, and the Standard chronometer is generally placed in the middle. Chronometers are made to run for 1, 2, or 8 days, but most generally for 2 days: thus of 44 chronometers sent for trial in 1880, 43 were 2 days, and only a single instrument was made to run for 8 days. An 8-day chronometer is usually wound every 7 days, and a 2-day chronometer is wound every day. The winding ought to take place at the same hour every day for the following reasons.

(1) To insure punctuality and constant habit, so that memory may not prove treacherous.

(2) To insure an even daily rate, for 24 hours, not 20 hours one day, 30 hours the next.

(3) To insure the *same part of the chain being always in use*, because, if the fusee is not accurately cut, we may obtain a different arc of vibration of the balance.

241. The winding is performed by a certain number of half-turns of the key, and this number ought to be exactly known and recorded on a slip of paper attached to the inside of the cover, so that in case of absence or sickness another officer might without fear be entrusted with the duty. In winding, gently invert the chronometer, draw back the spring-plate at the bottom, and insert the key; then turn firmly and evenly from right to left, as in an English lever watch, i.e. in the direction contrary to the motion of the hands. As the last half-turn is reached, (the chronometer must always be wound as far as possible), more care ought to be exercised, lest a sudden jerk might cause injury. It may, however, be noted that the resistance to further winding is produced, not by the end of the chain, but by a catch provided to act at the proper moment, and thus to save the chain from being strained or broken.

When a chronometer has stopped, it does not start again by itself on being wound up. When the winding has been completed, take the case into the hands, and give it a moderately quick and firm circular motion through an arc of about 90 degrees. It is found that if the time of stoppage does not exceed two or three days, the chronometer generally resumes its former daily rate*.

242. COMPARISON OF THE CHRONOMETERS.

When a chronometer is sent for service on board ship its error on G.M.T. at noon on the day on which it is removed, and its daily rate, are sent with it. Of course its performance on board is soon ascertained, and if its rate has changed under its new conditions, this new rate is known as the "harbour rate." When the ship has made a voyage, its error is determined at the port of arrival, either by single or equal altitudes, and the difference between the error at starting and that now determined, when divided by the number of days which have elapsed, gives the "sea rate" or "travelling rate." It seldom happens that the sea and harbour rates are exactly the same. In the rates of chronometers, it is observed that a constant acceleration takes place, i.e. losing rates decrease, and gaining rates increase. This result is due principally to three causes, (1) infiltration of dust, (2) thickening of the oils, (3) wear of the pivots.

* Here again there seems to be a difference of opinion, but if the instrument does not stop for more than the above time, it may be assumed that its rate is the same as before.

243. Method of Comparing.

The Standard is generally denoted by the letter Z, and the other chronometers by the letters A, B, C, ..., or by their numbers, or makers' names. It is found *convenient*, for the purposes of comparison, to consider the Standard as *fast* on all the other instruments, adding 12 hours to the Standard's time when necessary. Most chronometers beat half seconds, but some beat 5 times in 2 seconds; hence the former will beat 10 times in 5 seconds, and the latter 10 times in 4 seconds.

First, to compare the chronometers A, B, C with a deck watch M. Place M where its beats can be distinctly heard, and where its face and those of the chronometers can be seen.

Write down the *hour* and *minute* of M at which the comparison is to be made, e.g. let $M = 3^h\ 27^m\ 00^s$. Then, when the seconds hand of M arrives at 55^s (if M beats $\frac{1}{2}$ seconds), or at 56^s (if M beats 5 times in 2 seconds), count 0, at the following beats count 1, 2, 3,...; and at the 6th or 7th beat, cast the eye quickly on the face of the chronometer, and at the 10th beat, note carefully whether the chronometer's seconds hand is (1) at an exact second or $\frac{1}{2}$ second, or (2) passing from the second to the half second, or (3) from the half second to the second, and estimate as follows:—

In the first case write $0^s\cdot0$ or $0^s\cdot5$, in the second case write $0^s\cdot2$ or $0^s\cdot3$, in the third case write $0^s\cdot7$ or $0^s\cdot8$.

Suppose A marks $6^h\ 14^m\ 8^s\cdot3$, then $M - A = 15^h\ 27^m\ 0^s\cdot0 - 6^h\ 14^m\ 8^s\cdot3$ (12 hours added to M) $= 9^h\ 12^m\ 51^s\cdot7$.

After a minute, compare B in like manner. Suppose $B = 7^h\ 23^m\ 9^s\cdot7$. Subtract 1 minute, and write $7^h\ 22^m\ 9^s\cdot7$, then $M - B = 15^h\ 27^m\ 0^s\cdot0 - 7^h\ 22^m\ 9^s\cdot7 = 8^h\ 4^m\ 50^s\cdot3$.

After another minute, compare C. Suppose $C = 4^h\ 37^m\ 51^s\cdot3$, subtract 2 minutes and write $4^h\ 35^m\ 51^s\cdot3$, then $M - C = 15^h\ 27^m\ 0^s\cdot0 - 4^h\ 35^m\ 51^s\cdot3 = 10^h\ 51^m\ 8^s\cdot7$.

In this way all the comparisons *are reduced to the same hour*. If Z is the standard, then the others are compared with it.

The method of comparison on board ship is merely to compare the other chronometers with the Standard. This may be done in the following manner:—

Write down the time shewn by the Standard to some exact minute. The Standard most commonly beats half seconds. At 55" count 0, then 1, 2, 3, ... at the next successive beats; at the sixth or seventh look at the seconds hand of the chronometer to be compared (suppose, Dent, or Frodsham, or Eiffe, or A, or some number), and at the tenth beat note the part of the decimal according to the method just explained. Write down these seconds and decimals of a second, next the minutes, and lastly the hour, and take the difference; the comparison is then effected.

244. Copy of a Chronometer Journal (Shadwell, p. 29).

Date.	$Z-A$	2nd Diff.	$Z-B$	2nd Diff.	$Z-C$	2nd Diff.	Remarks.
1st	3 24 53		6 55 29		1 4 6		Heavy Gale.
2nd	3 24 54·7	1·7	6 55 41·2	12·2	1 4 4·5	1·5	Confused Cross Sea.
3rd	3 24 55	0·3	6 55 52	10·8	1 4 3·7	0·8	Less Motion.
4th	3 24 55·8	0·8	6 56 2	10·0	1 4 2·7	1·0	Night Quarters.

Remarks on the above extract:

(1) If the daily rates of Z and A are the same, then $Z-A$ is always the same.

(2) If the daily rates of Z and A are not the same, but if the rates are uniform, then the Second Difference is the same.

(3) If the Second Difference is found to vary, it is evident that either Z or A is altering.

To determine a faulty chronometer we must introduce a third instrument.

Thus we have $(Z-A)-(Z-B) = B-A$, or $A-B$. Here Z has been eliminated. Now on comparing the daily values of $Z-A$, and $A-B$, if it is found that the Second Difference of $A-B$ remains constant, while that of $Z-A$ is irregular, we infer that A and B are going steadily, and Z is altering; but if the Second Difference of $Z-A$ and $A-B$ are both irregular, while that of $Z-B$ is uniform, then A is the guilty one, while Z and B are going satisfactorily*.

When several chronometers are to be regulated by single or equal altitudes, these observations are made on shore by the aid of a Deck Watch which must be compared with the Standard before going on shore and on returning, and then the other chronometers must be compared with the Standard. The double comparison will eliminate any difference in the rates of the Deck Watch and the Standard.

Example. The following comparisons were made before and after observations.

	Before.			After.		
	h	m	s	h	m	s
Standard =	8	17	0·0	8	47	0·0
D. Watch =	10	8	9·5	10	38	8·0
	10	8	50·5	10	8	52·0

Deck Watch shews $10^h\ 19^m\ 13\cdot3^s$ when the observations were made, required the chronometer time of the observations.

The Deck Watch lost $1^s\cdot5$ in 30 minutes, ∴ in 11 minutes (viz. the difference between the first comparison and time of observation) it lost $\frac{11}{30}$ of $1^s\cdot5$ or $0^s\cdot55$, ∴ difference between Standard and Deck Watch at moment of observation = $10^h\ 8^m\ 51^s$.

Deck Watch at moment of observation = $10^h\ 19^m\ 13^s\cdot3$
Difference between Standard and Deck Watch at this moment.............. } = 10   8   51
∴ Standard's time at moment of obs. = 8   28   4·3

Remember that the Standard is *fast* on all the other chronometers.

* Shadwell, p. 31.

EXAMPLES FOR EXERCISE.

(1) The following comparisons were made between the Standard and a Deck Watch before and after an observation. Before, Standard $9^h 16^m 0^s \cdot 0$, D.W. $2^h 37^m 18^s \cdot 5$. After, Standard $9^h 29^m 0^s \cdot 0$, D.W. $2^h 50^m 20^s \cdot 5$. The time of observation by the D.W. was $2^h 46^m 15^s \cdot 3$; find the Standard's time of observation.

Result, $9^h 24^m 55^s \cdot 4$.

(2) The following comparisons were made before and after an observation.

Before, Standard $3^h 27^m 0^s \cdot 0$, D.W. $11^h 14^m 16^s \cdot 5$.
After, Standard 3 38 0·0, D.W. 11 25 20.

The time by D.W. when the observation was made was
$11^h 20^m 56^s \cdot 5$.

Required the Chronometer time of observation.

Result, $3^h 33^m 37^s \cdot 9$.

(3) The following comparisons were made:

Before, Standard $5^h 14^m 30^s$, D.W. $3^h 11^m 16^s$.
After, Standard 6 24 0·0, D.W. 4 20 48·7.

Time by D.W. when the observation was made $3^h 56^m 19^s \cdot 5$.
Required time of observation by the Standard.

Result, $5^h 59^m 31^s \cdot 8$.

245. In order to estimate correctly the interval $0^s \cdot 2$ of a second, the following method has been suggested. Stand in front of a wall which returns an echo, and about 110 feet from it: any sharp noise, such as clapping the hands, will have its echo returned in $\frac{2}{10}$ second. A little practice will render most persons quite competent to determine this portion of time by the ear.

246. COMPARISON OF A CHRONOMETER WITH A SIDEREAL CLOCK.

When two chronometers beating $\frac{1}{2}$ seconds are compared, it will seldom happen that their beats are coincident; they differ by a fraction of a second, the amount of which, by the methods already described, must be estimated by the ear. It is different, however, in the case where a chronometer and a sidereal clock are compared. In this case, it is quite possible to estimate the difference between the times by the two instruments within $\frac{1}{20}$ second, or, by practice, within even a smaller fraction. Since 1ˢ sidereal time is less than 1ˢ mean time, the beats of the clock will *gain* on those of the chronometer, and certain beats will exactly coincide. If the comparison can be made *at this instant*, of course the error of the one instrument on the other is determined. The only difficulty, therefore, arises from the impossibility of distinguishing this exact instant; but it is found that the ear will detect the non-coincidence of beats as long as the beats differ by $\frac{1}{20}$ second, and hence the comparison may be obtained within that amount.

Now 1ˢ sidereal time = 0ˢ·997 mean time, ∴ a sidereal clock gains 0ˢ·003 on the chronometer in 1ˢ, and ∴ gains $\frac{1}{2}$ second in something less than 3 minutes. About every three minutes, therefore, the two instruments will have coincident beats, and when this is about to occur, the observer begins to count the beats of the chronometer while he looks at the clock: when his ear can no longer detect any difference between the beats, he notes the corresponding seconds of the two instruments, then writes down the minutes and hours, and the comparison is made.

Example. A chronometer and sidereal clock were compared by coincident beats as follows:

	First obs.			Second obs.		
Chron.	4ʰ	16ᵐ	0·0ˢ	4ʰ	19ᵐ	10ˢ
Clock	1	3	11·5	1	6	22
Diff.	3	12	48·5	3	12	48

EXAMINATION.

(1) In what does the excellence of a chronometer over an ordinary time-piece consist?

(2) Explain the principle of the "gridiron pendulum."

CHRONOMETERS.

(3) Who first introduced with success the method of finding the longitude by means of chronometers?

(4) But a previous attempt had been made; and a still earlier suggestion made on the subject?

(5) Can you give any description of the mechanism which regulates the movement of a chronometer?

(6) Explain the principle of the *fusee*.

(7) Describe in some detail the balance of a chronometer, specifying the position of the metals, and noting the effect of changes of temperature.

(8) Mention the two chief methods of "compensating for temperature."

(9) Draw a diagram to illustrate Airy's compensation. Explain its object, and its special advantage.

(10) What is the "heat test" to which chronometers are exposed in the Royal Observatory?

(11) What grave objection is there to interfering on board ship with the position of the weights or screws of the balance?

(12) Explain the construction of a Chronometrical Thermometer, and the use of such an instrument.

(13) What is meant by the "Trial Number" of a chronometer? and write down the formula by which it is obtained.

(14) If a chronometer is sent by land, what precautions must be taken in the packing?

(15) Mention the precautions which ought to be observed in transporting chronometers from shore to ship.

(16) Specify the most favourable conditions for the situation of a "Chronometer Room."

(17) What are the two chief evils to be guarded against in the case of chronometers? Give reasons for your answer.

(18) How would you stow the chronometers in their places?

(19) What is the general tendency of variation of temperature on the going of a chronometer?

(20) Specify the two chief sources of error in the rates of a chronometer.

(21) Notice the general effect of lapse of time on the rate of a chronometer.

(22) Name the causes of this.

(23) Give the *three reasons* why a chronometer ought to be wound at the same hour daily. Which is the most important of the three?

(24) Describe the method of winding a chronometer.

(25) If a chronometer runs down, how is it set going again?

(26) Distinguish between the "harbour rate" and the "sea rate." How are they determined?

(27) How would you make a comparison between a Deck Watch and a Chronometer, on the supposition that the Chronometer beat half-seconds, and the Deck Watch beat five times in two seconds?

(28) If you were in charge of the chronometers in a ship describe your method of comparing them daily.

(29) What is the use of the "Second Difference" column in the Chronometer Journal?

(30) You have three chronometers on board, and reasons having arisen for doubting their performances, describe how you would detect the least trustworthy instrument.

(31) How are a chronometer and sidereal clock compared? Within what probable error will the comparison be effected?

CHAPTER XIII.

MERIDIAN DISTANCES.

247. DEF. THE Meridian Distance between two places *is measured by the exact difference of time* at the two places. Hence if a chronometer which shews the exact mean time at a place A be carried to a place B, and its error on B be determined, we can compute the difference of the times at A and B, and therefore the Meridian Distance of A and B.

The Meridian Distance is generally expressed in *time*, the difference of Longitude in *arc*.

248. DEF. Meridians are either primary or secondary. A *Primary* meridian is that from which different nations begin to count their longitude. E.g. England, the United States, Sweden, Norway, Prussia, and Austria reckon from the meridian of Greenwich; France reckons from Paris; Spain from Cadiz; Portugal from Lisbon, &c. A *Secondary* meridian is the meridian of a place whose longitude from a Primary meridian is known exactly. Thus on the shores of the North and South Atlantic Oceans there are 19 such Secondary meridians recognized: 7 in the Indian Ocean and Red Sea, 7 in the Eastern Seas, 9 in the Southern Ocean, and 8 in the Pacific.

249. Their use may be thus explained. A ship wishes to run a Meridian Distance with the object of finding the difference of longitude between, we will suppose, St John's, Newfoundland,

and Halifax, Nova Scotia. She finds, by her chronometers, that the difference of time at the two places is $0^h\ 43^m\ 37^s\cdot 5$, and that Halifax is to the westward of St John's. Now St John's is a Secondary meridian and its longitude is known to be 52° 40' 47" W.; if therefore to this the above Meridian Distance expressed in arc (= 10° 54' 23") be applied we obtain 63° 35' 10" W. the longitude of Halifax.

It may be also noted that the Meridian Distance between two places A and B does not require the knowledge of the longitude of either place. The Meridian Distance applied to the longitude of one of the places afterwards found will give the absolute longitude of the other.

The surveyor therefore confines his attention to a most careful measurement of the Meridian Distance between some Secondary meridian and the place, the longitude of which it is desired to fix absolutely.

250. This Meridian Distance is measured in two ways:

(1) *By the Electric Telegraph*, where possible.

(2) *By Portable Chronometers.*

251. I. BY THE ELECTRIC TELEGRAPH. Of course where telegraphic communication exists between two places, the mean of a number of signals must give a very perfect result*. But there are very many places where, at present, the only possible method of measuring a Meridian Distance is by transporting chronometers, carefully manipulated, from one point to another. We proceed to describe this method.

252. II. BY PORTABLE CHRONOMETERS.

In running a Meridian Distance the following precautions ought to be observed:

---

* For the details of such a method, the reader is referred to the "Report on the Difference of Longitude between the U.S. Naval Observatory and the Sayre Observatory of Lehigh University," by Prof. Eastman, Washington, 1878. This Report is in the Library of the Royal Naval College.

(α) The chronometers ought to be stowed on board so that surrounding circumstances may have least influence on their performance.

(β) Never to be touched except for winding and comparisons.

(γ) Only one person to have access to them.

(δ) To be wound at the same hour daily.

(ε) The comparisons to be made with the "Standard," backward and forward.

(ζ) The error to be determined on mean time at place before starting, at equal intervals of time (5, 7, or 10 days), to ascertain the performance of individual chronometers.

(η) The error to be determined by equal altitudes in the following manner:

    i. Compare the Deck Watch with Standard before going ashore.

    ii. Observe in the forenoon thus:

        Take the altitude of the ⊙'s L.L.
        Take the   ,,   ,,   ⊙'s U.L.
        Set the index 2° on, and take L.L. and U.L.
        Take 6 sets, 3 with each eye.

    iii. On coming on board compare Deck Watch with the Standard.

    iv. Compare again before going ashore for the afternoon observations.

    v. Take plenty of time, and set index at last altitude, and repeat the observations in a reverse order.

    vi. Again compare the Deck Watch with the Standard on finally arriving on board.

        Vide *Naval Science*, Vol. III. p. 95 ff.

253. The following formula will enable us to find the approximate time by the Deck Watch when the P.M. observations ought to be taken.

238 MARINE SURVEYING. [CHAP.

Let $t =$ time shewn by watch at the last A.M. observation.

$x =$ error of watch on S. M. T. (supposed slow).

$e =$ equation of time, supposed to be subtractive from mean time.

$t + x =$ mean time of last A.M. observation.

∴ $t + x - e =$ apparent time of last A.M. observation.

∴ $12 - (t + x - e) =$ time to apparent noon, or the apparent time of the first P.M. observation.

∴ $12 - (t + x - e) + e =$ mean time nearly of first P.M. observation.

∴ $12 - (t + x - e) + e - x =$ time by watch of first P.M. observation.

∴ $12 - t - x + e + e - x =$ time by watch required.

$12 - t - 2x + 2e = 12 - t - 2(x - e) =$ time required.

E.g. Let watch shew at the last obs. in the forenoon, $5^h 7^m 20^s$, watch slow $4^h 12^m 25^s$ on S. M. T. Equation of time 8^m additive to mean time. Then we can find the approximate time by watch for the first P.M. observation as follows :—

```
Time by watch   5ʰ  7ᵐ
        Error   4  12 +
                ───────
      M. time   9  19
      Eq. time      8 +
                ───────
     App. time   9  27
                12
                ───────
                 2  33 = interval to noon
                      = P.M. app. time.
                 2  33
      Eq. time       8 −
                ───────
                 2 ·25 = P.M. mean time.
        Error   4  12 −
                ───────
                10  13 = time by watch for the first P.M.
                              observation.
                ───────
```

254. We can see that if the rate of a chronometer does not change in being carried from one place to another, in other words, if the "travelling rate" or "sea rate" is the same as the "harbour rate," it will be very easy to discover the difference of time at the two places. These rates, however, are very seldom the same, and hence the computation is somewhat more tedious.

Example*. Where the travelling rate is supposed to remain the same. At Greenwich, May 5th, at mean noon, a chronometer shewed $11^h\ 49^m\ 42^s\cdot75$, its rate being $2^s\cdot671$ gaining. On May 17th, at a place A in the United States, at mean noon, it shewed $4^h\ 34^m\ 47^s\cdot28$.

Required the longitude of the place A.

$$\text{May } 4 \quad 23^h\ 49^m\ 42^s\cdot75$$
$$\text{May } 17 \quad\ \ 4\ \ \ 34\ \ \ 47\cdot28$$

Elapsed time $= 12\ \ 4\ \ 45\ \ \ 4\cdot53 = 12\cdot198$ days.

Accumulated rate $= 2\cdot671 \times 12\cdot198 = 32\cdot58$ seconds.

$$\text{Time at } A = 4^h\ 34^m\ 47^s\cdot28$$
$$\text{Acc}^d.\text{ rate } = \qquad\quad 32\cdot58 -$$
$$\overline{\qquad\qquad\qquad 4\ \ \ 34\ \ \ 14\cdot70}$$
Time at Greenwich $11\ \ 49\ \ 42\cdot75$

Long. in time $4\ \ 44\ \ 31\cdot95 = 71°\ 8'\ 00''$ W.

255. It may be remarked, for the sake of the younger readers, that since the Sun, in his daily course, arrives at the meridian of a place A to the eastward of Greenwich *sooner* than it arrives at the meridian of Greenwich, the time at A is *fast* on Greenwich time.

E.g. If a watch which shews the time at Bombay is brought to Malta, it is fast on Malta time; if brought to Gibraltar, it will be still more fast on mean time at place.

Again, suppose that a watch which shews Greenwich mean time were taken to Rome, it will be *slow* on the mean time at place; it will be still more slow on time at Constantinople, &c.

* From Chauvenet.

Once more, if a watch which keeps Greenwich time is found to be slow at a place A, and still more slow at a place B, we infer that A and B are both in *East* longitude, and that B is *farther East* than A.

Finally, if a watch which shews mean time at A is found to be fast on mean time at B, then it is evident that B is to the Westward of A.

The following questions ought to be thoroughly mastered before the solution of the Meridian Distances is attempted.

EXAMINATION.

(1) A chronometer is slow $3^h\ 12^m\ 17^s$ on A, and is slow $3^h\ 24^m\ 16^s$ on B; is A east or west of B?   A *west of* B.

(2) A chronometer is fast $7^h\ 16^m\ 5^s$ on A, and slow $1^h\ 14^m\ 18^s$ on B; is B east or west of A?   B *east of* A.

(3) A chronometer is fast $2^h\ 14^m\ 8^s$ on A, and fast $4^h\ 17^m\ 58^s$ on B; is A east or west of B?   A *is east of* B.

(4) A chronometer is slow on A, $5^h\ 27^m\ 33^s$, and fast on B, $2^h\ 18^m\ 49^s$; is B east or west of A?   B *west of* A.

(5) A chronometer is slow on A, $3^h\ 15^m\ 16^s$; ought the chronometer to be fast or slow on B, in order that the place whose time the chronometer shews may be midway between A and B?
    Fast on B.

(6) The time at B is fast on a chronometer $1^h\ 16^m\ 35^s$; the chronometer is fast on A, $5^h\ 11^m\ 51^s$. Is A to the east or west of B?   A *is west of* B.

256. It may be well to give here the usual Formulæ by which the "sea rate" of a Chronometer is found, when running a Meridian Distance between two places A and B.

Let a_1 = error of Chron. at A before sailing,
       b_1 = ,,   ,, B on arrival,
       b_2 = ,,   ,, B before sailing,
       a_2 = ,,   ,, A on arrival.

XIII.] MERIDIAN DISTANCES. 241

Then $(a_1 \sim a_2)$ = total error accumulated during the absence of the Ship from A (Harbour and Sea Error).

and $(b_1 \sim b_2)$ = total error accumulated during the stay at B (Harbour Error).

$\therefore (a_1 \sim a_2) - (b_1 \sim b_2)$ = amount of Error accumulated *at Sea*, in the two runs.

Let m = number of days from A to B,

and n = number of days from B to A.

$$\therefore \text{Sea rate} = \frac{(a_1 \sim a_2) - (b_1 \sim b_2)}{m + n}.$$

If only *one error* is determined at B, then we must use the less satisfactory Formula

Sea rate = $\dfrac{a_1 \sim a_2}{x}$, where x is the number of days the ship is absent from A.

Then, knowing the sea rate, we can compute at any moment the Error of the Chronometer on the time at A.

Finally, *Meridian Distance* = Error of *Chron.* on time at A ~ *Error of Chron.* on time at B.

Meridian Distance worked out:—

(For greater refinements in the methods of computation, the reader may consult Shadwell on Chronometers, Ed. 1861, Chapter VIII.).

Example 1*.

August 1, chron. fast on Cape Passaro time 7^h 54^m $5^s\cdot 7$
  ,, 2,   ,,   Messina   ,, 7 52 23·7
  ,, 8,   ,,   Messina   ,, 7 52 19·2
  ,, 9,   ,,   Cape Passaro ,, 7 54 0·6

---

* Examples 1, 2, 3 are taken from Papers in Naval Science on Nautical Surveying by Capt. Shortland, Vol. III., p. 101. See Capt. Shortland's remarks, p. 102.

Compute the Meridian Distance between Cape Passaro and Messina.

$$\begin{array}{ll} \text{Aug. 1, } 7^h\ 54^m\ 5^s\cdot7 & \text{Aug. 2, } 7^h\ 52^m\ 23^s\cdot7 \\ \text{,, 9, } 7\ \ 54\ \ 0\cdot6 & \text{,, 8, } 7\ \ 52\ \ 19\cdot2 \end{array}$$

Loss in 8 days   5·1   Loss in 6 days   4·5
(harbour and sea).   (harbour).

                                          Loss in 8 days   5·1
                                          (harbour and sea).

$\therefore$ loss during the 2 days at sea   0·6
                or sea rate =   0·3 −

August 1, Chron. fast on Cape Passaro $7^h\ 54^m\ 5^s\cdot7$
                   Change in one day at sea   0·3 −

$\therefore$ August 2, Chron. fast on Cape Passaro 7 54 5·4
    ,,   ,,   Messina 7 52 23·7
         $\therefore$ Meridian Distance = 0 . 1 41·7

Again, August 8, Chron. fast on Messina 7 52 19·2
             Change in one day at sea   0·3

$\therefore$ August 9, Chron. fast on Messina 7 52 18·9
    ,,   ,,   Cape Passaro 7 54 0·6
      $\therefore$ Meridian Distance = 0 1 41·7

Since chronometer is faster on Passaro than on Messina, it follows (by Ex. 3 in the Examination) that Passaro is to the westward of Messina.

Example 2.

Sept. 4th, Chron. fast on Le Have mean time $5^h\ 35^m\ 26^s\cdot17$
  ,, 5th,   ,,   Prospect   ,,   5 33 13·47
  ,, 10th,   ,,   Prospect   ,,   5 33 27·25
  ,, 11th,   ,,   Le Have   ,,   5 35 45·34

Compute the Meridian Distance.

Result, $2^m\ 15^s\cdot39$ *Le Have to the westward of Prospect.*

Example 3.

Sept. 5th, chron. fast on Prospect mean time 5ʰ 33ᵐ 13ˢ·47
„ 6th, „ Halifax „ 5 32 34·66
„ 9th, „ Halifax „ 5 32 42·70
„ 10th, „ Prospect „ 5 33 27·25

Compute the Meridian Distance.

Result, 0ᵐ 41ˢ·68 *Prospect to the westward of Halifax.*

Example 4. On leaving a port A, in longitude 118° 4' E. a chronometer was slow on S. M. T. 7ʰ 47ᵐ 9ˢ·6; daily rate 1ˢ·9 *gaining;* on arriving at B, the chronometer was slow on S. M. T. 7ʰ 55ᵐ 55ˢ·3; daily rate 2ˢ·03 *gaining;* the interval between these observations was 2·99 days. Required longitude of B.

(Exam. Papers, May, 1876.)

Mean rate = 1·965 gaining.

Accumulated rate in 2·99 days = 5ˢ·86 gained.

Chron. slow on A 7ʰ 47ᵐ   9ˢ·6
Accum. rate   5·86 gained.
∴ Chron. slow on A after 2·99 days 7 47 3·74
Chron. slow on B at same time   7 55 55·30

∴ Meridian Distance = 0 8 51·56; and B is to the
east of A.

D. long. = 2° 12' 53" E.
Long. of A = 118 4 0 E.

∴ Long. of B = 120 16 53 E.

Example 5. On leaving Malta, long. 14° 31' E., chron. was fast on S.M.T. 1ʰ 20ᵐ 11ˢ·5; daily rate 7ˢ·23 *gaining;* seven days afterwards on arriving at Alexandria, long. 29° 51' 40" E., the chron. was fast on S.M.T. 0ʰ 20ᵐ 27ˢ·55. Find what change had taken place in the daily rate. (Exam. Papers, Oct. 1877.)

244   MARINE SURVEYING.   [CHAP.

$$\begin{array}{rl} \text{Long. Alexandria} =& 29°\ 51'\ 40''\ \text{E.} \\ \text{,, Malta} =& 14\ 31\ 00\ \text{E.} \\ \hline \text{D. long.} =& 15\ 20\ 40 \\ \text{In time} =& 1^h\ 1^m\ 22^s\cdot 6 \\ \text{Chron. fast on Malta} & 1\ 20\ 11\cdot 5 \\ \hline \end{array}$$

∴ If chron. had no daily rate, it ought to be fast on Alexandria }   0 18 48·9

$$\begin{array}{rl} \text{It actually was fast} & 0\ 20\ 27\cdot 55 \\ \hline \therefore \text{Accum. rate in 7 days} =& 0\ \ 1\ \ 38\cdot 65 \text{ gained} \\ \therefore \text{Daily rate} =& 0\ \ 0\ \ 14\cdot 09 \text{ gaining} \\ \text{Former rate} =& 0\ \ 0\ \ \ 7\cdot 23 \text{ gaining} \\ \hline \therefore \text{change in rate} =& 0\ \ 0\ \ \ 6\cdot 86 \text{ gained} \end{array}$$

Example 6. A chronometer was slow on Monte Video mean time $1^h\ 14^m\ 14^s$ (long. 56° 10′ W.); daily rate $1^s\cdot 8$ losing. Fourteen days afterwards at sea, the summit of Tristan d'Acunha in lat. 37° 17′ S., long. 12° 36′ W. was observed to bear N. 38° E. (true), distant 5 miles, when the chronometer was found to be slow on S.M.T. $4^h\ 9^m\ 00^s$. Find change in the rate.

(Exam. Papers, June, 1878.)

First to determine the exact position of the ship.

S. 38° W. 5′, ∴ D. lat. =   4′ S.   Dep. = 3′·1, ∴ D. long.
Lat. of island 37° 17 S.   (= 3·1 sec. 38° 19′)
∴ Lat. in = 37 21 S.   = 0° 3′ 57″ W.
                                            Long. of island = 12 36 00 W.
                                            ∴ long. in 12 39 57 W.

$$\begin{array}{rl} \text{Long. of Monte Video} =& 56°\ 10'\ \ 0''\ \text{W.} \\ \text{,, Ship} =& 12\ 39\ 57\ \text{W.} \\ \hline \therefore \text{D. Long. in arc} =& 43\ 30\ \ 3 \\ \text{,, in time} =& 2^h\ 54^m\ 00^s\ \text{ship to the east.} \\ \text{Chron. slow on Monte Video} & 1\ 14\ 14 \\ \hline \end{array}$$

$\therefore$ If Chron. had no rate it would have been slow on S.M.T. $\Big\}$ $4^h \ 8^m \ 14^s$

It actually was slow $\quad 4 \quad 9 \quad 00$

$\therefore$ Accum. rate in 14 days = $\quad 0 \quad 0 \quad 46$ lost.

$\therefore$ daily rate = $3^s \cdot 28$ losing
Former rate = $1 \cdot 80$ losing

$\therefore$ change in rate = $1 \cdot 48$ in a losing direction.

Examination, and Examples for Exercise.

(1) What is a Meridian Distance?

(2) Define the terms Primary and Secondary meridians, and mention the use to which the latter may be put.

(3) Describe the precautions which ought to be taken in running a Meridian Distance. (Dec. 1877.)

(4) Mention the two principal methods of measuring a Meridian Distance.

(5) Having found a Meridian Distance how would you ascertain in what direction it ought to be applied to fix the longitude of a station? Write down an example to illustrate your answer.

(6) When the mean time is 3^h at Greenwich it is $7^h \ 51^m \ 12^s$ at Bombay, and $13^h \ 4^m \ 56^s$ at Sydney. What is the Meridian Distance between Greenwich and Bombay, and between Bombay and Sydney? (Exam. Papers, Sept. 1877.)

Results. (1) $4^h \ 51^m \ 12^s$; (2) $5^h \ 13^m \ 44^s$.

(7) On leaving a port in longitude $116° \ 39' \ E.$, a chronometer was slow on S.M.T. $8^h \ 36^m \ 00^s$; what would be its error on S.M.T. in longitude $119° \ 30' \ E.$ (rate not considered)? (Nov. 1875.)

Result. $8^h \ 47^m \ 24^s$ slow.

(8) A foreign chart shews a rock in long. 9° 54' west of the meridian of St Petersburgh. The mean time of St Petersburgh is $2^h\ 1^m\ 19^s$ fast on Greenwich M.T. What longitude should be assigned to the rock on an Admiralty chart? (March, 1878.)

Result. 20° 25' 45" E.

(9) A Time Ball is dropped at an Observatory in lat. 34° S. at 1 P.M. local mean time; a chronometer on board a Ship 6 miles S.E. (true) of the Observatory, shews at the instant $1^h\ 29^m\ 16^s$. Required the error of the chronometer on S. M. T.

(April, 1880.)

Result. D. Long. between Ship and Obs. = 5' E. = 20^s in time. ∴ When ball dropped S.M.T. = $1^h\ 0^m\ 20^s$, ∴ error of chronometer = $0^h\ 28^m\ 56^s$ *fast.*

(10) A Time Ball drops at 1 P.M. mean time of Melbourne Observatory. A chronometer shewed $1^h\ 30^m\ 30^s$ at the same instant. The Observation Spot in lat. 39° S. bears west (true) 3·4 miles from the Observatory. Required the error of the chronometer on mean time of the Observation Spot. (Nov. 1877.)

Result. $0^h\ 30^m\ 47^s$ *fast.*

(11) The last A.M. observation of a set of equal altitudes was taken at $12^h\ 47^m\ 34^s$ by a chronometer slow on A.T. at place $8^h\ 43^m\ 48^s$. Find the approximate time by the chronometer for the first P.M. observation. (April, 1876.)

Result. $5^h\ 44^m\ 50^s$.

(12) The last A.M. observation of a set of equal altitudes was taken at $1^h\ 51^m\ 20^s$ by a chronometer slow on A.T. at place $7^h\ 39^m\ 00^s$; what would be the approximate time by the chronometer for the first P.M. observation? (Oct. 1876.)

Result. $6^h\ 50^m\ 40^s$.

(13) The last A.M. observation of a set of equal altitudes was taken at $6^h\ 14^m\ 19^s$ by a deck watch slow on S.M.T. $2^h\ 21^m\ 11^s$; equation of time $7^m\ 36^s$ subtractive from mean time; what would be the time by the watch for the first of the P.M. observations?

Result. $1^h\ 18^m\ 31^s$.

(14) Compute the Meridian Distance between Trincomalee and Madras from the following errors of chronometers:

June 23 slow on Trincomalee M.T. $5^h\ 39^m\ 52^s\cdot82$
,, 29 ,, Madras ,, 5 36 39·30
July 14 ,, Trincomalee ,, 5 42 15·19
(May, 1881.)

Result. Trincomalee $0^h\ 3^m\ 54^s\cdot38$ east of Madras.

(15) On leaving a port A, in longitude 120° 16′ E. a chronometer was slow on M.T. at place $8^h\ 7^m\ 30^s$, daily rate $2^s\cdot23$ gaining; and on arriving at B, chronometer was slow on S.M.T. $7^h\ 42^m\ 58^s\cdot3$, daily rate $2^s\cdot15$ gaining; the interval between the observations was 7·02 days. Required the longitude of B.

(Sept. 1876.)

Result. Long. $B = 114°\ 11′\ 55″$ E.

(16) Aug. 5th a chron. was fast on Malta M.T. $2^h\ 39^m\ 30^s\cdot5$,
,, 12th ,, ,, 2 39 45·9.
Sailed on Aug. 12 for Gibraltar.
Aug. 20th the chron. was fast on Gib. M.T. 3 59 12·5,
,, 26th ,, ,, 3 59 27·5.
Compute the Meridian Distance between Malta and Gibraltar.

(Feb. 1878.)

Result. Malta is $1^h\ 19^m\ 7^s\cdot8$ E. of Gibraltar.

(17) On leaving a port A, in long. 114° 11′ E. a chronometer was slow on S.M.T. $4^h\ 35^m\ 2^s\cdot9$; daily rate $12^s\cdot16$ gaining. On arriving at a harbour B, the chron. was slow on S.M.T. $4^h\ 59^m\ 7^s\cdot5$; daily rate $12^s\cdot88$ gaining; the interval between the observations was 14·98 days. Required the longitude of B.

(March, 1877.)

Result. Long. of $B = 120°\ 59′$ E.

(18) On August 1st a chronometer was fast on Malta M.T. $2^h\ 13^m\ 13^s$, daily rate gaining $1^s\cdot56$; on August 7th the chron. was fast on Cyprus M.T. $0^h\ 57^m\ 51^s$, daily rate $1^s\cdot85$ gaining. Find the Meridian Distance between Malta and Cyprus, using the mean rate.

(Sept. 1878.)

Result. Cyprus is $1^h\ 15^m\ 32^s\cdot23$ E. of Malta.

248   MARINE SURVEYING.   [CHAP.

(19) A chronometer was slow on S.M.T. (long. 56° 20' W.) $1^h 14^m 29^s$, daily rate $1^s \cdot 8$ losing; 15 days afterwards at sea, an island, in Lat. 37°10' S., Long. 12° 39' W. bore N. 25° E. (true) distant 4 miles, when the chronometer was slow on S.M.T. $4^h 9^m 10^s$. What change had occurred in the rate?   (Aug. 1880.)

Result. Long. ship $= 12° 41' 8''$ W. . Change $1^s \cdot 44$ in a gaining direction.

(20) On leaving Portsmouth, in Long. 1° 6' 15" W. at noon, a chronometer was fast on M.T. at place $3^h 5^m 2^s$; daily rate gaining $1^s \cdot 28$. Find the time by chronometer at 1 P.M. Plymouth M.T. on the succeeding day, the Longitude of Plymouth being 4°10' 15" W.
                                                                 (Oct. 1879.)

Result. 1 P.M. at Plymouth $= 1^h 12^m 16^s$ at Portsmouth.
∴ Rate must be computed for 1·05 days.
∴ Chron. shewed $4^h 17^m 19^s$.

(21) Left Sydney on July 10th, a chronometer being fast on S.M.T. $2^h 10^m 0^s \cdot 45$, rate $3^s \cdot 5$ *gaining*. July 19th, the chronometer was fast on S.M.T. at Tanna Island $3^h 22^m 59^s$. July 31st, on returning to Sydney the chronometer was fast on S.M.T. $2^h 11^m 3^s$. Calculate the Meridian Distance between Sydney and Tanna Island.   (Oct. 1877.)

Result. Tanna Island is $1^h 12^m 31^s \cdot 7$ W. of Sydney.

Note. The given rate is not required in the solution.

(22) On March 3rd in Harbour:

Chron. *A* was *fast* at noon on S.M.T. $0^h 4^m 40^s$, rate $3^s \cdot 25$ *gaining*.
„   *B* was *slow*   „   „   7 52 50, rate $0 \cdot 33$ *losing*.

On March 10th at Island:

Chron. *A* was *fast* at noon on S.M.T. $1^h 4^m 41^s$, rate $3^s \cdot 18$ *gaining*.
„   *B* was *slow*   „   „   6 52 50, rate $0 \cdot 15$ *gaining*.

Compute the Meridian Distance between the Harbour and the Island by each chronometer, and state which result you would prefer, and your reason.   (March, 1879.)

Result. By *A*, Island west of Harbour $0^h 59^m 38^s \cdot 53$.
By *B*   „   „ $1^h 0^m 0^s$.

XIII.]   MERIDIAN DISTANCES.   249

The result by A to be preferred, the rate of A being more uniform.

Additional Questions from the Examination Papers set for Classes B_1 and B_2.

(23) On leaving A, in Longitude 122° 13' E. chronometer X was slow on M.T. at place $7^h 54^m 27^s \cdot 5$, daily rate $3^s \cdot 70$ gaining; at B, X was slow $8^h 14^m 20^s \cdot 8$; and at C, X was slow $8^h 24^m 16^s \cdot 7$, daily rate $3^s \cdot 35$ gaining. The interval between the observations at A and B was 5·99 days, and between the observations at B and C 5·99 days. Required the Longitudes of B and C.

(June, 1876.)

Results. Long. of $B = 127° 16' 36''$ E.; Long. of C $= 129° 50' 51''$ E.

(24) On leaving A, in Longitude 114° 11' E., chronometer was slow on M.T. at place $7^h 16^m 23^s \cdot 10$, daily rate $3^s \cdot 24$ gaining; at B, chronometer was slow $7^h 42^m 54^s \cdot 95$, daily rate $2^s \cdot 04$ gaining; the interval between the observations was 14·98 days. Required the Longitude of B.

(June, 1877.)

Results. Long. of $B = 120° 58' 45''$ E.

(25) May 8th, chronometer fast on M.T. at Mauritius, $3^h 45^m 46^s$, daily rate gaining $2^s \cdot 85$.

The same chronometer fast on M.T. at Aden $4^h 36^m 21^s \cdot 75$ at noon on May 19th, and at noon of May 26 fast on M.T. at Aden $4^h 36^m 46^s \cdot 25$.

Ascertain the Aden rate; also the Meridian Distance between Mauritius and Aden, using the mean of the rates.

(June, 1878.)

Results. Aden rate $= 3^s \cdot 5$ gaining.

Meridian Distance $= 0^h 50^m 0^s \cdot 83$ Mauritius to the east of Aden.

(26) June 1st, chron. fast on M.T. Colombo $\quad 3^h 48^m 47^s \cdot 5$
,, 8 ,, ,, Singapore 2 12 59·5
,, 15 ,, ,, Singapore 2 13 20·8
,, 29 ,, ,, Hong Kong 1 32 49·4

Find the Singapore rate, and by means of it determine the meridian distance between Colombo and Hong Kong.

(June, 1879.)

Results. Singapore rate = 3^s·04 gaining.

Meridian Distance = 2^h 17^m 23^s·2 Hong Kong to the eastward of Colombo.

(27) At noon July 6th, a chronometer slow on M.T. Singapore 7^h 44^m 15^s·4; on proceeding at once to Malacca, the error of the same chronometer on M.T. of that place, July 12th (noon) was 7^h 38^m 40^s·9 slow, and after remaining at anchor, the error at noon of July 16th was 7^h 39^m 8^s·7 slow; the return voyage to Singapore was then made, and on July 21st (noon), at that latter place the error was found to be 7^h 46^m 12^s·7 slow; required the Meridian Distance between Singapore and Malacca.

(June, 1880.)

Result. Meridian Distance = 0^h 6^m 23^s·3 Singapore to the eastward of Malacca.

CHAPTER XIV.

METHOD OF PLOTTING A SURVEY.

257. Having in the preceding chapters described with more or less detail, as seemed necessary, the different methods of constructing and using scales, laying off angles, fixing positions, measuring base lines, conducting a triangulation, taking tidal observations, running lines of soundings, managing chronometers and computing meridian distances, it seems well in this last chapter to gather into one compass the various steps in a Survey of a harbour.

258. The operations then may be arranged under two headings.

I. The method of setting about the work, and the data to be obtained.

II. The method of projecting on paper the details obtained in I.

Projecting the work on paper is technically known as Plotting.

259. I. The method of setting about the work.

(*a*) Certain remarkable features are fixed on, and suitable names assigned by which they will be known, such as West mound, Hut station, Flag station, Cairn, &c.

(*b*) Select the best position for the Base Line.

(*c*) Measure the length of the Base.

(*d*) Determine the Lat. and Long. of one extremity. This point is known as the Observatory Station.

The *Latitude* will probably be determined by the meridian altitude of the sun and stars, observed in the artificial horizon.

The *Longitude* by sun chronometer, the sun's altitude being observed in an artificial horizon.

(*e*) The *Direction* of the base will also be accurately determined.

(*f*) From the Observatory Station, angles must be taken with the Theodolite to *all* the important points in the survey which are visible. Similarly angles must be taken from all the other stations. And a closer series of angles from such stations as are suitable to enable the coast line to be cut in.

(*g*) The tidal observations are also to be made, to determine the Establishment of the Port, Rise and Range of tide, &c.

(*h*) The Sounding operations are to be carefully made and reduced to the datum line.

(*i*) The Levelling work must also be attended to, for the sake of sections, contours, &c.

(*j*) Topographical details, features of the coast, &c. are to be noted.

260. II. Plotting the work.

Def. We may define this operation as the construction of a figure similar to the projection of all the remarkable points of the survey on a plane, a tangent to the Earth's surface, at the middle point of the locality.

The paper ought to be mounted on a drawing-board. The scale ought to be drawn on the paper, and will therefore vary as the paper itself is subject to the changes in the atmosphere.

If the plotting is to be engraved it is essential that the scale should be engraved with it, because the paper just before receiving the impression is damped and consequently expands. In a length of 2 feet the error will be $\frac{1}{4}$ or $\frac{3}{8}$ of an inch, a difference which would render a plan almost useless. Indeed it is seldom that any two impressions from the same plate are of exactly the same size, owing to the moisture in the paper at the time of printing. Printing on dry paper is never found to give satisfactory results*.

It is often convenient in Plotting Work to have the True or Magnetic Bearings of the several sides of the triangles. This information is secured by the process known as "Surveying *by the Back Angle.*" At the First Station, the Bearing of the Zero is found (§ 127), thus all angles observed may be treated as Bearings, and these may be continued throughout the entire chain by setting the Instrument at each station to the angle from the last one †.

In important surveys the triangulation must be put in by the method described in §§ 45, 46. This method enables us to lay off angles to single seconds. The coast line is generally put in by the aid of a rectangular protractor.

The following directions for Plotting the Questions in Examination Papers will, it is hoped, be of some use to the Younger Students.

(1) In a convenient position on the paper make a dot, and inclose it in a small circle; this will represent the Observatory Station.

(2) From this point draw the Base Line on the proposed scale of the plan and make a clear mark at the other extremity of the Base.

(3) Lay off the angles observed at the Observatory Station and at the other end of the Base. These "cuts" will "fix" the principal objects as seen from the extremities of the Base Line.

(4) Then proceed to "fix" the other points or stations by means of the angles taken at the various stations.

(5) Afterwards cut in the coast line by the angles observed for that purpose.

---

* Maxton's *Engineering Drawing*, p. 233.
† Jeffers, p. 122.

254   MARINE SURVEYING.   [CHAP.

(6) Insert all the topographical information and finish off the frame, &c.

We shall now plot one or two of the Examples taken from the different Examination Papers, and add a copious collection from the same source for exercise.

261. Suppose then that we are required to plot the Beaufort Paper for November, 1878 (No. 8 in the following Exercises):—we proceed as follows.

(1) Compute the Sun's True Bearing from the known latitude, declination and altitude of the centre. This is found to be N. 74° 30′ E.

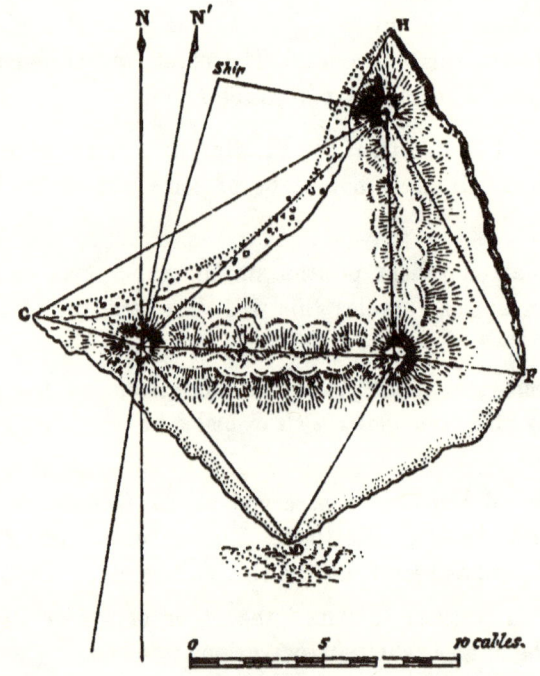

(2) Next compute the length of the base line from the known angles of depression to the mainmast and the height of the maintruck 130 feet. Thus we find the base line to be 6070 feet, which expressed in the given scale will be 2·01 inches.

XIV.]   METHOD OF PLOTTING A SURVEY.   255

(3) We place a dot in the middle of the paper and rule the true meridian through it; and then lay off N. 74° 30′ E. This will be the direction of the sun.

(4) The *direction* of C will evidently be 28° 30′ *to the left of this line*. From this line we can lay off the angles observed at A.

(5) Lay off the direction of the ship 29° to the left, or 331° to the right.

(6) Lay off 2·01 inches found in (2)*. This gives the position of the ship.

(7) Knowing the angle at C (vide the data) between the zero A and the mainmast to be 54° 40′, and also knowing the angle at A between the zero and mainmast 29° 0′, we can evidently find the third angle of the triangle. This will be 96° 20′. At ship lay off this angle, and the line will cut the zero line in the position of the station C. Hence we have fixed three points in the Survey.

(8) We next lay off 45° 15′ to the right of the zero from A. This will pass through B. We then lay off to the left of zero from C an angle of 48° 15′ (360° − 311° 45′). These lines will intersect in the station B.

(9) G is fixed by the angles from A and C.
   H .... A ... C.
   D .... A ... B.
   F .... B ... C.
   Boulder .... A ... C.

(10) The angles subtended at ship by the points C and H, A and G respectively, will be found to correspond with those observed by the sextant on board. Thus proving that these points are correctly fixed.

(11) The Topography of the island will depend on the draughtsman's fancy, but all the directions given must be carefully attended to. A knowledge of the various Symbols and Conventional Signs is requisite in putting in the given details.

* The plans are reduced owing to the size of the present pages.

(12) We next take the distances *AB*, *BC*, *CA* from the diagonal scale in $\frac{1}{2}$ inches, and thus compute the lengths in *miles*.

(13) The Variation is found to be 9° 50 E; accordingly the True and Magnetic Meridians can be ruled through *A* as directed.

(14) The height of observer's eye at *A* above the horizontal plane through the maintruck is 181 feet. Hence the total height of *A* above the sea is $181 + 130 - 5 = 306$ feet.

(15) We draw a line two inches long to represent a mile. We divide this into 10 equal parts. Each will represent a cable. We next draw another line a short distance from the first and parallel to it and finish off according to the directions given in the chapter on Scales (§ 17).

262. Suppose that we have to plot the Final Paper set June, 1878 (No. 15 in the following Exercises), we proceed as follows:

(1) Compute the sun's true azimuth. This is found from the latitude, altitude and declination to be N. 74° 40' E.

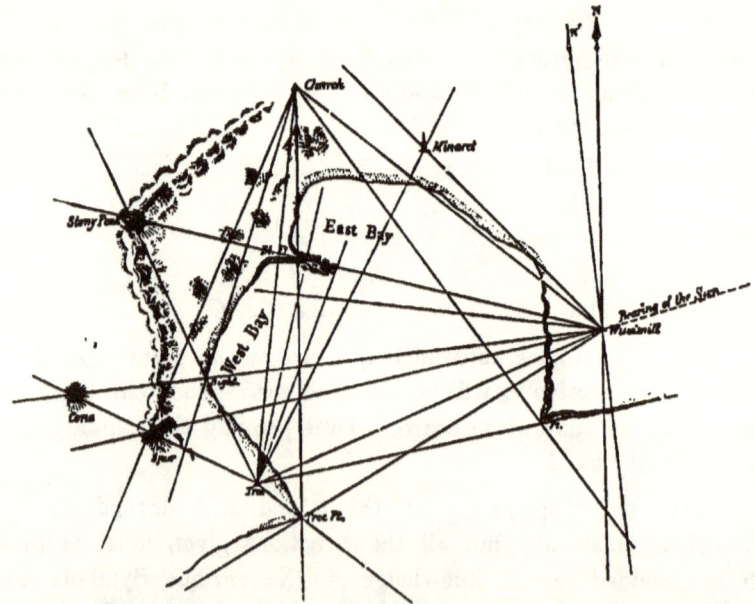

(2) The distance between the Windmill and Church △'s is found to be represented by 4·5 inches.

METHOD OF PLOTTING A SURVEY.

(3) A spot is selected about 4 inches from the right-hand edge, and some distance from the lower edge. This will be the Windmill station.

(4) Draw a line to represent the True meridian. Lay off 74° 40' *to the right*, and then 50° 57' to the *left of the meridian*. This will be the *direction* of the Church △. Along this line take off 4·5 inches. This will give the exact position of Church △.

(5) From Church with Windmill as Zero, lay off 57° 30', this will be the direction of Tree △. From Windmill with Church as Zero lay off 61° 10' (= 360° − 298° 50'). These lines will intersect in the position of the Tree △.

(6) We next fix the Minaret by the angles from Windmill and Tree, and *test* the "fix" by the angle from Church.

We next fix Tree Point, from Windmill and Church, and test by the angle from Tree △.

Then Spur is fixed, from Windmill and Church, and tested by the given angle from Tree.

&c. &c.

As each point is fixed, it is well to put a small dot in *ink* and write its name in pencil near it.

When all the points have been plotted, we must read the topographical information very carefully, and sketch lightly *in pencil* the outlines required, and then with a crow-quill and Indian ink put in the work as neatly as possible.

The following very valuable example is, by Sir E. J. Reed's permission, taken from *Naval Science*, Vol. I. p. 342.

At Observatory △ (White Rock).
South hill 58° 0' ship's foremast.
North hill⎫
φ depth of Sandy bay⎭ 43° 45'   „   „
Hill over Rocky point 31° 00'   „   „
South hill 42° 20' Coast mound.
Ship's foremast 78° 00' Flat rock.

Flat rock 63° 10′ out ext. Cliff h^d.
" " 70° 00′ Cliff ends, small bay begins.
P^t inside obs^y △ 89° 30′ hill over Rocky p^t.
⟶ Sandy p^t 44° 00′ " " "
⟵ Cliff beyond,⎫ 34° 20′ " " "
near Sandy bay⎭
Fisherman's hut (northern) 33° 10′ " " "
South hill 15° 5′ Stake in depth of large bay.
" " 22° 25′ Sandy bay ends.
" " 25° 5′ near part of next Point.
" " 28° 55′ inner extreme of Rocky Ledge.
" " 32° 10′ {outer ditto,
    depth of Sandy bay beyond ϕ.
" " 45° 20′ ⟶ Coast Mound p^t.

At Ship's Foremast.

Obs^y △ 77° 10′ South hill.
South hill 35° 10′ North hill.
" " 42° 30′ Hill over Rky P^t.
Hill over Rocky p^t 34° 00′ Coast Mound.
Flat r^k 14° 15′ Obs^y △.
Obs. △ 1° 25′ Point inside Obs^y △.
" 5° 20′ ⟶ p^t. Beach begins.
" 9° 20′ Sandy p^t.
" 16° 15′ Depth of small bay.
" 28° 50′ ⟵ Cliff
" 32° 30′ nearest part of Cliff.
" 36° 20′ ⟶ Cliff. Beach recommences.
" 43° 14′ Southern hut.
" 46° 00′ Middle "
" 48° 00′ Northern "

Bay runs ½ cable outside the huts, in a gentle curve to Stake in its depth 93° to the right of Obs^y △.

South Hill 31° 00′ Beach ends.

XIV.]   METHOD OF PLOTTING A SURVEY.   259

Point looking this way just right of the North Hill, and ⟵ of Rocky Ledge φ Hill over that point.

South Hill 48° 30' {extreme of Rocky Ledge.
                    {Beach beyond commences.
„   „   59° 50' Depth of northern bay.
„   „   73° 00' Beach ends.
Hill over Rocky p^t 34° 30' Coast Mound.
„   „   „   37° 50' ⟶ Coast Mound P^t.
„   „   „   43° 10' Rock awash.

Angles at South Hill Station.

Ship's Foremast 44° 50' Obs^y △.
North Hill 98° 15' Ship's foremast.
Hill over Rocky p^t 59° 50' „   „
Ship's foremast 45° 15' extreme of Point inside Obs^y △.
„   „   47° 20' Sandy p^t.
„   „   51° 00' Centre of small cliff.
„   „   55° 20' Southern hut.
„   „   56° 00' Middle   „
„   „   54° 00' Northern „

P^t inside Rocky p^t } 55° 15' Ship's Foremast.
φ North Hill from ship }
Rocky ledge begins 54° 10'   „   „
Outer extreme of do. 48° 20'   „   „

Angles at Hill over Rocky point Station.

Ship 77° 50' South Hill.
Coast Mound 100° 00' Ship's Foremast.
Obs^y △   7° 30' Depth of small bay, right of Obs^y △.
Ship   φ   ⟶ Rocky ledge.
Coast Mound 61° 00' ⟵ Rocky ledge.
„   „   47° 15' Rock awash.
Depth of Northern bay 117° 00' Ship's Foremast.
⟶ Coast Mound p^t 93° 00'   „   „
Sandy bay ends φ Coast Mound △

17—2

Angles at Coast Mound Station.

Ship 45° 00′ South Hill
„ 46° 00′ Hill over Rocky point.
„ 78° 00′ North Hill.
„ 42° 00′ Rocky patch begins.
„ 33° 00′ Outer extreme of do.
„ 53° 40′ {Sandy bay begins and curves gently round thence to termination.
Rock awash 58° 00′ Hill over Rocky point.

Angles on Flat rock.

← Cliff head 90° 00′ Obsʸ △.
→ do. Southern bay begins 99° 5′ do.
Depth of Small bay 23° 20′ do.

To plot the above work:—

(1) Make a dot on the most convenient part of the paper, and enclose it in a small circle. This will represent the Obsʸ △.

(2) Through this Obsʸ △ draw a *true* N. and S. line.

(3) Lay off the T.B. of the ship at the other end of the Base, which in this case is N. 26° 50′ E.; and the length of the Base = 2·05 miles, determined by Sound and Masthead angle.

(4) Lay off 58° 50′ the angle at Obsʸ △ between Ship and South Hill, to the left of the ship. Then 79° 10′ at the Ship between South Hill and the Obsʸ △. This fixes South Hill. Now as the angle between Obsʸ △ and the Ship observed at South Hill is 44° 50′, the work is correct.

(5) Project in the same way the triangle formed by Ship, South Hill, and Hill over Rocky point. Then the triangle formed by Ship, Hill over Rocky point, and Coast Mound. All these points were used as Stations.

(6) But North Hill was not used as a station, and is "fixed" by the angles obtained at the Ship between it and South Hill, and that obtained at South Hill between the Ship and it; and is proved by a line of direction obtained at the Obsʸ △ between it

XIV.] METHOD OF PLOTTING A SURVEY. 261

and South Hill passing through the point of intersection of the lines of direction obtained from the Ship and South Hill.

Vide diagram I. for this triangulation, and diagram II. for the cutting in of the coast line.

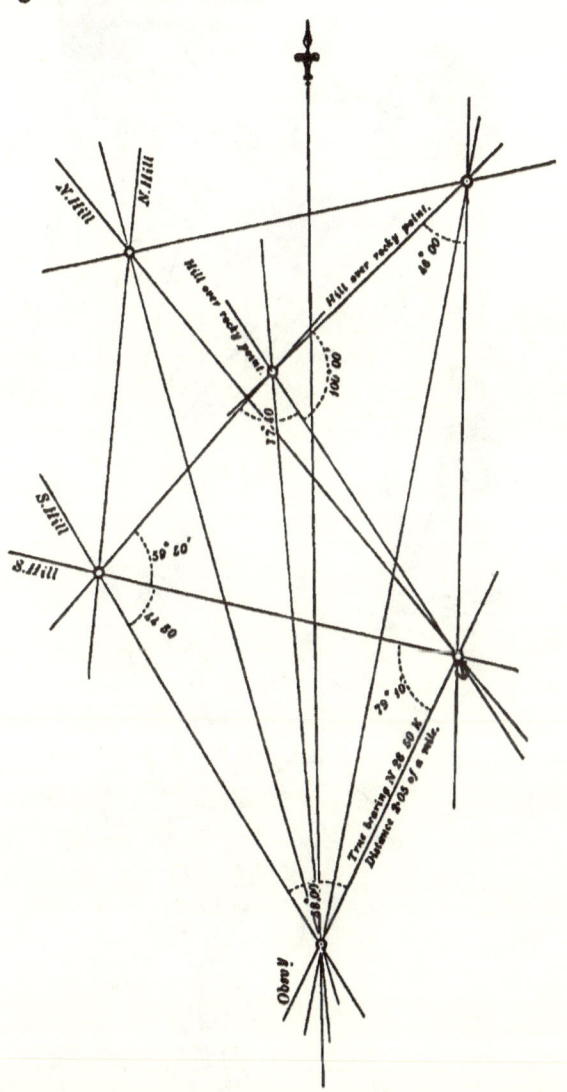

Diagram I. Shewing the method of plotting the principal triangles, and fixing the chief points.

262 MARINE SURVEYING. [CHAP.

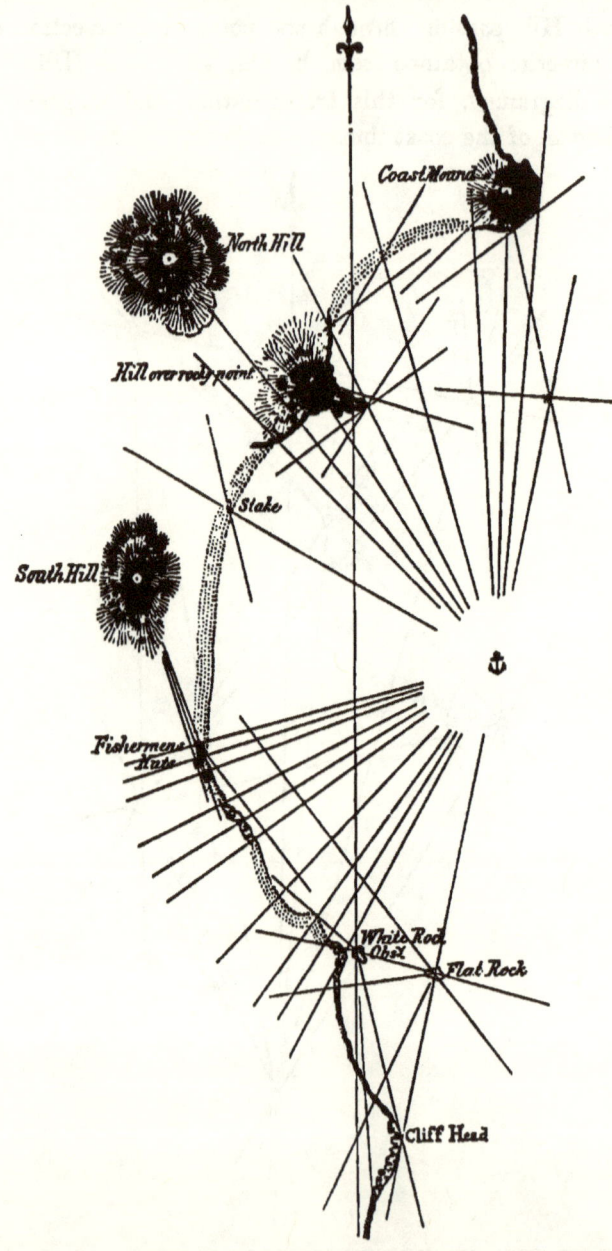

Diagram II. Shewing the method of cutting in the coast line, and representing the various features.

BEAUFORT PAPERS.

No. 1. September, 1874.

In the exploration of a coast, a lofty peak, Mount Columbus, known to be in lat. 54° 52' N., long. 160° 18' E., was observed from the ship to be directly in line with a hill near the shore on a *true* bearing of N. 60° W. At the same time the angle between Mt Columbus and a Peaked island to the south-west was found to be 80°.

The vessel then stood to the south-west until the Peaked island came in line with Mount Columbus on a *true* bearing of N. 36° W.; at the same time the angle between the Hill near the shore and the Peaked island was 25°, and that between the Peaked island and a Cliff farther to the south-west, known as Cape Drake, was 101°. The altitude of Mount Columbus was here observed; on the arc 0° 23' 50", off 0° 22' 40"; height of eye 18 feet.

The vessel proceeding to the south-west, when off Cape Drake a landing was effected, and Cape Drake found from observation to be in lat. 54° 16' 30" N., the *true* bearing of Mt Columbus being N. 16° W., and the angle between the Mount and the Peaked island was found to be 32° 30'.

Project these positions on a scale of an inch to a mile, and determine *by projection* the longitude of Cape Drake, and the *true* bearing and distance between that Cape and the Hill near the shore; also calculate the height of Mount Columbus.

Results. Long. Cape Drake 160° 20' 36" E.
Hill near shore bears from Cape Drake N. 12° E., 26·6 miles.
Mt Columbus is distant 32 miles.

No. 2. April, 1875.

Arriving at a coral reef, and there being too much swell to attempt a landing, a position was taken up westward of it with Gull rock (on the west side of the reef, and at the northern edge of an entrance to a shallow lagoon), in line with a single Palm (on a small sand bay near the eastern extreme) bearing N.69°E., and the following angles observed:—

A. N.W. extreme of reef 41° 0′ Gull rock 11° 20′ lagoon entrance,
,, ,, 49° 0′ S.W. extreme,
course N.ᵇE. for 0·98 miles, to

B. North extreme 77° 10′ Gull rock 8° 0′ W. and S.W. extreme in line,
course N.E.ᵇE. 1·06 miles, to

C. N.E. extreme 81°0′ Gull rock 26°0′ N.W. extreme
Palm 70° 0′ ,, ,,
course E.ᵇS.½S. 1·67 miles to

D. S.E. extreme 44°0′, Palm in line with lagoon entrance 42°0′N.E. extreme,
course S.ᵇW. 1·6 miles, to

E. South extreme 81° 0′, Palm 13° 0′ East extreme,
course W.ᵇS. 2·05 miles, and the vessel then anchored.

F. West extreme 16° 0′ Gull rock 10° 0′, lagoon entrance,
,, ,, 28° 0′, Palm,
,, ,, 53° 0′, South extreme.

Observations placed the ship in latitude 10° 31′N., longitude 115° 7′ E., and gave the *true* bearing of the Palm N. 52° E.

The courses and bearings are cor. mag. and the variation 5° E.

Project these positions on a scale of 2 inches to a mile, and determine latitude and longitude of the Palm, giving sketch of reef thus roughly surveyed.

Results. Lat. Palm = 10° 32½′ N.
Long. ,, = 115° 9′ E.
Distance = 4·87 inches, or 2·43 miles.

No. 3. *October*, 1875.

In the survey of a coast, Bluff △ was found to bear N. 8°30′ E.

XIV.]   METHOD OF PLOTTING A SURVEY.   265

(true), 4·1 miles from South point △; the following △s were observed at:—

(1) South point △.

Zero, Bluff △	360° 00′
Rock awash	6 45
Round point	19 00
Wharf φ trend of coast line }	29 00
⟵ village	37 00
⟶ ditto	40 00
Rock hill △	52 20
This point out	267 00
Direction of Long reef, extending from this point }	332 00
⟵ Bluff reef	343 30
⟵ Bluff	358 00

(2) Bluff △

Zero, South point	360° 00′
⟶ South point	3 10
⟶ Long reef	55 00
Direction of Bluff reef extending from, and φ this point out }	71 30
Trend of coast northward }	165 00
Coast line in	296 00
⟵ Village	300 00
⟶ ditto	303 30
Round point	306 20
Wharf	315 45
Rock hill △	319 20
Rock awash	352 50

(3) Rock hill △.

Zero, Bluff △	360° 0′
Wharf	5 20
Round point	9 10
South point △	273 00
Rock awash	325 20
⟶ Long reef	330 30
⟵ Bluff reef	343 40

Both Long and Bluff reefs are coral, about 2½ cables in width, with patches above water; the coast line (sand backed with low bush) joins between the points that are cut in. Project this work on a scale of 1·52 inches to a mile, giving a sketch of the port.

Results. Distance between Rock awash and Rock Hill 2·94 in.
    South Hill 3·23 ,,
    Left of Bluff reef 3·20 ,,
    Extreme of Bluff and Long reef 1·38 ,,

No. 4. March, 1877.

In the survey of a port, lat. 23° 0' N., a base of 2000 feet was measured along its western sand beach, and the following △s were observed at:—

(1) South base △.

Zero, Signal hill	360°	0'
Mean of A.M. observations for the T.B. T.A. ⊖ 30° 31' }	85	15 ⌽
⟶ of Sand and ⟵ of Cliff }	2	0
⟶ of Cliff and ⟵ of Sand }	13	0
Rock on Sand beach	25	0
Bluff	39	40
⟵ Quail islet	56	00
Summit ,,	68	20
⟶ ,,	90	30
South point	116	20
Coast line near	137	00
North base △ and line of beach }	342	30

(2) North base △.

Zero, Quail islet summit }	360°	0'
⟶ do.	11	45
South point	28	00
South base △	48	30
Beach near	258	00
Signal hill	262	20
⟶ of Sand, and ⟵ of Cliff }	275	40
⟶ of Cliff, and ⟵ of Sand }	285	00
Rock on beach	297	50
Bluff	314	20
⟵ Quail islet	350	00

(3) Quail islet summit.

Zero, North base	360°	0'
⟶ of Sand ⟵ of Cliff	30	0
Signal hill	34	0
⟶ of Cliff, and ⟵ of Sand	46	0
Rock on beach ⌽ North end this islet }	62	20
Bluff	82	30
South extreme this islet	237	0
South point	265	20
South base	314	20

XIV.]   METHOD OF PLOTTING A SURVEY.   267

(4) Sextant △. At Rock on beach.

⟵ Quail islet 5° 0' Summit Quail islet 3° 50' ⟶ Quail islet,
Beach near 50° 30'   ,,   ,,   98° 0'   Signal hill
                                                     φ beach near.

The coast line joins between the points that are cut in; Quail islet, the extremes of Bluff and South point, and cliff fronting Signal hill are low and rocky, rest of beach sand: a reef awash about half a cable wide joins South point and the nearest part of Quail islet; sun's decl. 10° 45' N. Project and give a sketch of this work on a scale of 5 in. = 1 mile = 6055 feet.

Results.   T.B. of Sun N. 90° 51' E.
             S. P^t to Bluff 3·4 inches.
             S. △ to Signal hill 3·29 in.
             N. △ to Quail islet 2·29 in.
             S. △ to Rock on beach 3·54 in.

No. 5.   *May,* 1877.

In the survey of a large lagoon, latitude 20° 0' S., a base of 2300 feet was measured along the ridge of a sand islet forming its south-west side, and the following △s were observed at:—

(1) South Base △.   (2) North Base △.

Zero, Flag △	360° 0'		Zero, Flag △	360° 0'
Mean of obs. for T.B. taken at 6 A.M. A.T. } 102 30 φ			⟶ Tree islet	64 0
			⟵ This islet	76 0
⟶ Tree islet	56 30		South base △	102 30
⟵ This islet	66 0		North point this islet } 318 0	
Trend of do. near	90 0		φ centre of joining reef	
North base	314 30		Sentry rock	330 0
Sentry rock	342 30			

(3) Flag △. North extreme of Tree islet.

Zero, South base △	360° 0'		⟵ This islet	300° 0'
North base △	32 0		Bight in of do.	326 0
⟶ South-west islet	38 0		⟶ do.	328 20
Sentry rock	113 0		⟵ South-west islet	339 0

(4) Sextant △. Sentry rock at North extreme of reef.

Trend of reef near, then trends to Flag △ } 65° 0' S. Base △ 21° 30' N. ext. S.-w. islet
„ 40° 0' { Trend of reef near, then trends to S.-w. islet.

(5) Sextant △. South extreme Tree islet.

Near point, South-west islet 111° 0' Flag △ 5° 0' Bight in this islet in 35 0 ⟶ do. near.

Tree islet is covered with the cocoa-nut trees, about a cable across near the centre, and is joined to South-west islet, which is half a cable in breadth, by reefs above water of the same width; lagoon is shallow; Sun's declination 23° 0' S. Project and give a sketch of this work on a scale of 5 inches = one mile = 6053 feet.

Results. Sentry rock to N. base 2·75 inches.
Flag to S. base 3·50 inches.
N. pt. this islet to S. pt. Tree islet 3·07 inches.

No. 6. October, 1877.

In surveying a reach of river, ship anchored near mid-channel, and following observations were taken, base being measured by Sound.

(1) At B △, zero, A △, bearing N. 5° W. (*true*) and N. 26° W. (*Mag*₁).
Beats of chron. between flash and report of gun fired at ship ☉ 18½
House on hill   9° 0'
East point north shore   61 10
Prom. point south shore 78 58
Far point south shore   271 15
West point north shore
  φ ship ☉   306 0

(2) At A △, zero, distant peak.
Beats of watch between flash and report fired at ship ☉ 29
B △   22° 15'
Ship ☉   74 15
Far point south shore   80 10
West point north shore   105 0
House on hill   250 0
East point north shore   305 12
Prom. point south sh.   345 0
Sand point φ distant peak.

(3) At ship ☉, zero, A △.

East point north shore	31° 45'
Prom. point south shore	53 00
Sand point	59 30
B △	74 00
Far point south shore	191 0

Chronometer used at B △ has 2 beats to a second. Watch used at A △ has 5 beats to two seconds. Sound travels 1090 feet a second. Project these positions, scale 1·5 inches to mile = 6069 feet, using B's measurement of base. State what number of beats of *watch* A was wrong. Rule true and magnetic meridians. South shore of river is low and sandy, extending nearly straight from Far point past B △ to Sand point,—thence, in a small bay to Prominent point. North shore of river is straight between points named and A △, and consists of a low cliff, A △ being highest part. House is on a hill rising gradually above A △. Current W.ᵇS. 1·5 knots. Ship ☉ is in 9 fathoms, and a rock awash was found with

Ship ☉ φ West point 84° 0' A △.

Sketch in these details, and place the rock awash on the plan, giving magnetic bearing of house on hill from it.

Results. Watch at A in error 5 beats too many.
Mag. bearing of House from Rock awash N. 10° E.
Distance of A from Prom. point 3·27 inches.
„ West point from Far point 2·86 inches.
„ House from East point 3·52 inches.

No. 7. September, 1878.

In a running survey of an island, ship steamed on following courses, angles being taken at the position started from, and at the termination of each course:—

270   MARINE SURVEYING.   [CHAP.

Mag. Course.   *Dist.*
(1)   N. 78° E.   1'·25
(2)   N. 2° W.   1'·45
(3)   N. 55° W.   1'·3
(4)   S. 40° W.   2'·6

At Starting Position.

Lat. 0° 40' S., long. 10° E.

⟵ island 52° S. peak (N. 10° E.) elevated $\left.\begin{array}{c} \\ 0° 52' 23'' \end{array}\right\}$ 48° S.E. point.

At end of Course (1).

⟵ island 20° 30' S. peak ϕ S.E. point (N. 62° W.) $\Big\}$ 32° N. point.

N. peak 12° ditto

At end of Course (2).

S.E. point   61° 30' N. peak (S. 79° W.) 24° N. point.
S. peak   28° 30' ditto.

At end of Course (3).

N. point   17° 30' N. peak (south) elevated 0° 47' $\Big\}$ 26° island ⟶

⟵ island 19°   ditto.

Bearings magnetic. Variation 32° W. Projection plane.

Protract the above on scale of 2 ins. = a mile of 6046 feet, giving the height of N. and S. peaks, and lat. and long. of the former.

Measure the horizontal angle the island should subtend at the end of Course (4), and the distance round the coast.

Ground plan of island resembles an isosceles triangle. In approaching from the north or south the island appears of conical shape, but from the east and west is like a saddle, N. and S. peaks representing the horns. Shores consist of shingle beach.

Sketch in topography and detail.

XIV.]   METHOD OF PLOTTING A SURVEY.   271

Results:—

Height N. peak = 80·7 feet (dist. from 4th position 5638 feet).
  „   S.   „   = 78·7 feet („   1st   „   5169 feet).
Lat. and Long. N. peak 0° 37′ 18″ S. and 9° 58′ 30″ E.
Horizontal angle = 40°.
Distance round coast 4·62 miles.

No. 8.   November, 1878.

In survey of an island, △s were made upon three peaks A, B, C—base line being measured by height of a ship's main truck, moored in the offing.

At A △.   Lat. 20° N.
Height of eye 5 feet.

Zero, C △	360° 0′	Mag. bearing N. 36° E.
Sun's centre	28 37	⎧Corrected declination 18° 15′
„ „ alt. of	11 20	⎩ N. for A.M. true bearing.
Beach boulder	6 30	
B △	45 15	
Point D	95 30	
Point G	241 0	
Ship's mainmast	331 0	
„ main truck depressed	1 42 30″	⎱ For base line.
„ water line „	2 56	⎰
Point H	353 50	

At B △.

Zero, C △	360° 0′
Point F	99 0
Point D	211 5
⎧ A △	273 30
⎩ ϕ direction of range towards west	
Ship's mainmast	331 0

At C △.

Zero, A △	360°	0'
Point G	13	20
Ship's mainmast	54	40
Point H	142	0
Point F	285	30
B △	311	45
Beach boulder	353	0

Sextant angles from ship.

Point H 23° 30' C △.

A △   21 30 point G.

Ship's main truck to water line 130 feet.

Protract the above on scale of 2 inches = 6053 feet = nautical mile. Find the length of the sides of the triangle ABC. Rule true and mag. meridians through A △, and give its height above the sea surface.

Range of hills forms an elbow at B, extending to A and C. Beach boulder is on the coast, at the head of a rocky bay, which curves between points G and H. South portion of island consists of cultivated plain, with sand coast from point G to D; also from D to F. A tide rip off point D. Low cliff between points F and H.

Sketch in details, and divide the scale of the plan to cables.

Results. AB = 1·87 miles.
BC = 1·78 „
CA = 2·48 „
Var. = 9° 50' E.
A is 306 feet above the water.

No. 9. February, 1879.

In a survey of a group of islands a △ was made on the summit of High Island, ship mooring to the eastward, the cutter being sent round to take up a position to the westward of the group.

XIV.] METHOD OF PLOTTING A SURVEY. 273

High island △.   Height of eye 5 feet.
Zero, distant peak 360° 0' Mag. bearing N. 17½ E.
True bearing N. 27 E.

At Ship (Sextant Station).

⟵ A island	24	45				
Summit do.	31	00	⟵ High isl.	11° 5'	High island △	
⟶ do.	42	30	C rock	20 0	,,	
Ship's mainmast at			⟶ B island	24 55	,,	
water line	62	0	Summit do.	29 00	,,	
Do. depressed	2	2 45"	⟵ do.	32 50	,,	
⟵ B island	112	0			⟶ High isl.	
Summit do.	120	0	High isl. △	10 0	ɸ ⟵ of A and	
⟶ do.	128	30			D islands.	
C rock	181	30	Do.	18 30	A isl. summit.	
Cutter ☉	240	30	Do.	37 30	Do. ⟶	
⟵ D island	275	30				
Summit do.	292	30				
⟶ do.	312	00				

At Cutter (Sextant Station).

⟵ High island	14° 00'	High island △	
⟵ A island	15 30	,,	
⟶ D island	28 00	,,	
Summit do.	38 30	,,	
⟵ do.	45 30	,,	
High island △	16 30	⟶ High island.	
,,	20 30	⟵ B island.	
,,	25 30	Summit do.	
,,	30 00	⟶ do.	
,,	51 10	C rock.	

Distance by micrometer measurement between High island △ and ship's water line at mainmast 6068 feet.

Protract the above on scale of 2 inches = a nautical mile of 6060 feet.

Rule true and magnetic meridians, and give height of High island summit.

R. M. S.   18

The islands have steep cliffy coasts. High island is oval-shaped, greatest diameter north and south, with conical summit. C rock is a sharp pinnacle.

Sketch in detail; divide scale to cables, and find the *natural scale* of the plan.

Results. Var. 10° E.
High island is 194 feet above the sea.
Natural scale $=\frac{1}{36360}$.
$AB = 1\cdot 31$ inches, $CD = 1\cdot 53$ inches.
$BC = 1\cdot 08$,, $DA = 1\cdot 23$,,
D and ship $= 2\cdot 57$,,

No. 10. *September*, 1879.

In constructing a plan of Greenwich, including part of the river Thames, △s were made at the Observatory, Gasworks, and Chimney on Essex shore, the distance between the two former by calculation being 4137 feet.

Sun's declination 7° 40′ N.

At Observatory △ lat. 51° 28′ N.

Zero, St Paul's Cross	360 00′
Mag. bearing N. 36° 30′ W.	
Sun's centre	164 30
App. time VIII^h A.M. for true bearing.	
Ship flagstaff	10 50
Dogs Ry. Station	22 30
Obelisk φ Gymnasium	30 00
Church	34 30
Chimney △	43 30
Asylum	48 00
Gasworks △	343 40
Crane	356 20

At Gasworks △.

Zero, St Paul's Cross	360° 00′
Crane	51 30

METHOD OF PLOTTING A SURVEY.

Dogs Ry. Station φ Chimney △	98° 30'
Asylum	120 00
Ship flagstaff φ obelisk	128 00
Gymnasium	146 30
Observatory △	159 40

At Chimney △.

Zero, Observatory △	360° 00'
Gymnasium	4 20
Obelisk	13 52
Ship	33 00
Gasworks △ φ Dogs Ry. Station	59 00
Crane	72 00
Church	94 00
Asylum	352 00

Protract upon scale of 7 in. = a mile of 6,084 feet, ruling true and mag. meridian through Observatory △, and find the natural scale.

The river, 300 yards broad, curves (the south bank) from Gasworks △ close past Ship and Asylum; the north bank about parallel. Mark its course, state how and at what time of tide you would sound, and the manner in which the time of high water at full and change of moon should be observed.

Bearing in mind the features of the district, mark approximately where you would measure the base-line. Should the heights be given above, and soundings below the same level of river? State how you would fill in upon the plan the streets, buildings, and topography.

Results. Chimney to Obelisk = 2·48 in.
Asylum to Crane = 4·17 ,,
Station to Obser^y = 4·23 ,,
Gasworks to Church = 4·26 ,,
Variation = 18° 28' W.

No. 11. *October*, 1879.

In the survey of the extinct crater of St Paul's island (now a small harbour), a △ was made on the ridge at the head of the crater, and three ☉'s on the shores thereof.

At Ridge △.

{ Zero, Hut △
  φ Dip of ridge } 360° 0′

True bearing of zero S. 50° E.
Mag. bearing S. 66° 30′ E.

{ Eastern entrance ☉   30 0 }
{   ,,   ,,   depressed 4 30 }
West entrance point ☉   49 0
Red stone ☉   86 30
End of ridge on west side 91 40

At East entrance point ☉.

Hut ☉   29° 30′   Dip of ridge.
Ridge △   40 20   Hut ☉
Red stone ☉   85 00   ,,
End of ridge on } 97 10   ,,
   west side   }
West entrance } 113 20   ,,
   point ☉   }

At Hut ☉.

Red stone ☉   48° 0′   Ridge △.
End of ridge on } 53 0   ,,
   west side   }
West entrance pt. ☉   86 0   ,,
East   ,,   ☉   110 5   ,,

At Red Stone ☉.

East entrance point ☉   15° 50′ West entrance point ☉
Dip of ridge   23 35   East   ,,   ,,   ,,
Hut ☉   33 30   ,,   ,,   ,,   ,,
Ridge △   78 55   ,,   ,,   ,,   ,,

XIV.]   METHOD OF PLOTTING A SURVEY.   277

The distance between West entrance point and East entrance point ⊙, by line stretched across, was found to be 809 feet.

Protract the above upon scale of 7·4 in. = a mile of 6,060 feet.

Give the height of Ridge △, and rule the true and magnetic meridians through it.

The crater is circular, Ridge △ about 100 yards inland, the other ⊙ on the coast line, the entrance lying between the East and West entrance points. The ridge of the crater follows the curve of the coast, sloping gradually from the Dip of ridge on Eastern, and end of ridge on Western side, to East and West entrance points, otherwise sloping at an angle of 45° down to the water's edge.

Ridge △ seen midway between the points of entrance, is the leading mark into the harbour.

Sketch in these details, and give the magnetic bearing of the leading mark.

Results.   Height of ridge is 198 feet.   Bears N. 27° 30′ W.

No. 12.   *May*, 1880.

When making a plan of a harbour, a base 1,200 yards long was measured on a sandy beach on west side; angles as follows were observed at the north and south extremes of the base, and also from the summit of an island in the harbour:—

At South Base △ zero distant Peak.

N. Base △   15° 00′ (bearing N. 54° 30′ W. true).
Rock awash at H.W.   50 20
⟵ Island   83 50
Island △   88 30
⟶ Island ϕ east⎫   92 40
   entrance point⎭

At North Base △ zero S. Base △.

    ⟶ Island   314° 10'
    Island △   312 00
    ⟵ Island   309 00
    East entrance point   298 40
    Rock awash   287 30

At Island △ zero N. Base △.

Rock awash   23° 20' (angle of depression 1° 2').
East entrance point   133 00
S. Base △   301 30

From South Base △ the coast rises in cliffs and trends N. 70° E. (true) for 5 cables and then N. 13° W. (true) towards east entrance point, which is fringed by a coral reef.

On subsequently sounding the harbour, a shoal of 3 fathoms, coarse sand and coral, was discovered, and from which the following sextant angles were observed:

East entrance point 51° 30' Island △ 31° 00' S. Base △.

Protract the above on a scale of 6 in. = one mile (6,080 feet) and find height of the island.

No. 13. *June*, 1880.

From a vessel in lat. 48° 30' N., long. 28° 00' W., the following sextant angles were observed:—

    ⟵ island 23° 00' ⟵ beach
    ,,   32 15 Church
    ,,   43 30 summit of island ϕ ⟶ beach
    ,,   47 00 Windmill
    ,,   68 20 ⟶ island.

The vessel then steered S. 74° E. (true) for a distance of 2¼ miles, when the following angles were taken:—

    ⟵ island 7° 00' ⟵ beach
    ,,   19 00 Church ϕ ⟶ beach

XIV.] METHOD OF PLOTTING A SURVEY. 279

⟵ island 33° 10' summit of island
„   43 00   Windmill
„   57 00 ⟶ island

From this position the vessel was moved N. 21° E. (true) 3 miles, when the following angles were observed:—

⟵ island 40° 20' North point of island (distant 1¼ miles)
φ ⟶ island
„   28 40 summit of island

From the first position the ⟵ island bore N. 22° E. (magnetic), the variation being 10° W., and from the second position the same point bore N. 50° 30' W. (true). The island is triangular in shape; the north side is cliffy and nearly straight; the south-east extreme is surrounded by rocks which uncover at low water.

The following soundings were also obtained off south side of island:—

⟵ island 59° 30' Church 57° 20' Windmill (5 fms., hard sand). Church φ ⟶ beach 103° 00' ⟶ island (7 fms., mud and stones).

Protract the above on scale of 2 in. = a nautical mile.

Rule true and magnetic meridians through first position of vessel, and find lat. and long. of summit of island.

Results. Latitude of summit   48° 32' 11" N.
            Longitude   27 55 10 W.
            North point to south point   2·7 inches
            Left extreme to S.E. point   4·28 „
            North point to church   2·34 „
            Westward sounding to windmill 1·95 „
            Eastward   „   „   1·95 „

No. 14. *April,* 1881.

The following angles and bearings were observed from a ship, when running along a coast extending about 5 miles in a north and south direction:—

280   MARINE SURVEYING.   [CHAP.

From Ship anchored at A, in 12 fathoms, sand.

North islet (bearing N. 6° E. true) 15° 30' Yellow cliff.
,,   ,,   ,,   29 30 Rock awash ϕ head of Green bay.
,,   ,,   ,,   35 00 White cliff ϕ flagstaff on hill.
,,   ,,   ,,   60 30 Rock at head of White bay ϕ Hill one mile inland.
,,   ,,   ,,   88 00 South point of White bay.

Leaving the cutter anchored at A, the ship takes up a second position, B, in 9 fathoms, coral, at 3 miles N. 10° W. (true) from A, and observes as follows:—

From Ship anchored at B.

South point of White bay   22° 00' Cutter at A.
Rock at head of White bay   41 00   ,,
White cliff ϕ Hill one mile inland 56 30   ,,
Rock awash   69 30   ,,
Flagstaff on hill   94 00   ,,
Yellow cliff   113 00   ,,
North islet   142 30   ,,

The Cutter then weighs anchor and proceeds to C, a position $2\frac{1}{2}$ miles north (true) of the Ship at B, and anchors with Flagstaff on hill ϕ North islet.

From Cutter anchored at C in 7 fathoms, broken shells.

South point of White bay   19° 00' ship at B.
White cliff   33 00   ,,
Rock awash   38 30   ,,
Head of Green bay   57 00   ,,
Yellow cliff   66 15   ,,
Flagstaff on hill ϕ north islet 69 30   ,,

The coast forms two sandy bays (the southern known as

XIV.]   METHOD OF PLOTTING A SURVEY.   281

White bay and the northern as Green bay), separated by White cliff, a perpendicular cliff towards which the land slopes from the Hill one mile inland.

The hill on which Flagstaff stands falls steeply to the eastward, but slopes gently to Yellow cliff, the north point of Green bay.

North islet nearly joins the shore, and between it and Yellow cliff the coast forms a deep sandy bay.

Project the foregoing on a scale of one inch = one mile.

Note.—The position of A should be taken at 2 inches from the bottom of the paper and 3 inches from the left-hand margin.

                 Results. Distance B to N. islet 2·28 inches.
                          ,, B to W. cliff 2·50   ,,
                          ,, A to N. islet 5·10   ,,
                          ,, C to S. point 5·90   ,,

Note.—The following papers were set at the examinations of Classes B_1 and B_2.

No. 15. Final, 1878*.

In survey of a harbour △ s were selected at *Windmill* on table land on east side of entrance, at *Church* on hill about a mile inland from the harbour's head, and at *Tree* on a projecting headland on south-west side of entrance.

Working from a measured base line the distance between *Church* △ and *Windmill* △ was calculated to be 18,159 feet.

At *Windmill* △. Lat. 20° N.

Zero, *Church* △	360° 0'	Mag. bearing N. 45° W.
Sun's centre	125 37	⎰Corrected decl. 18° 5' N.
,, ,, altitude of	11 20	⎱ for A.M. true bearing.
Minaret φ end of sand in east bay	7 0	
Tree point	291 0	
Tree △	298 50	
Spur in tree range	310 0	

* See reduced Diagram in § 262.

Cone in tree range	316° 10'
South end of reef extending from Black point	329 0
Stony peak φ Black point	335 30

At Church △.

Zero, *Windmill*	360° 0'
Windmill point	15 · 0
South end of reef	44 0
Black point	48 0
Tree point	51 0
Tree △	57 30
Bight of West bay	69 0
Spur in tree range	75 10
Cone „ „	89 30
Stony peak „	103 30
Minaret	347 0
End of sand in east bay	357 50

At *Tree* △.

Zero, *Church* △	360° 0'
Black point	7 0
South end of reef	14 30
Minaret	20 30
End of sand in east bay	46 0
Windmill △	61 20
Windmill point	72 50
Tree point	120 0
Spur φ cone in tree range	290 0
Stony peak φ bight of sand in west bay	329 0

West bay (shore of sand) curves between Tree point and Black point; East bay (shore of sand) curves between Black point and end of sand. Black point is a rocky shelf dividing East from West bays. Reef of rocks awash extends southward from Black point. Coast between end of sand and Windmill point straight

XIV.]   METHOD OF PLOTTING A SURVEY.   283

and cliffy. Minaret surrounded by scattered houses. Cultivation around the head of the bay. Tree range of hills extends between *Church* and *Tree* △s, decreasing in height towards the latter.

Project and give a sketch of this work on a scale of 1·5 inches = a mile of 6,053 feet, ruling the true and magnetic meridians through *windmill* △.

N.B. *Windmill* △ *should be about* 3 *inches from the right-hand edge of the paper.*

  Results. Var. 5° 57′ W.
    Tree point to Windmill point 3·02 inches.
    Church △ to Tree △ 4·55 ,,
    Tree △ to Windmill 4·40 ,,
    Tree △ to Stony peak 3·33 ,,
    Black point to Windmill point 3·28 ,,

    No. 16. *March*, 1880.

In making the survey of a bay a base was obtained by calculation between Beacon △ and Signal hill △ equal to 7,500 feet, and the following angles taken :—

 Latitude of Beacon △ 50° N.: Sun's declination 20° N.

    At Beacon △.
    Magnet N. 20° E.

Zero, Signal hill △	360°	0′
⟶ Cliff	18	30
Bight of sand	28	20
Extreme of sand spit	34	50
Rock awash	41	10
Rock on beach	43	20
Bluff △	60	0
⟶ Bluff	63	30
Extreme of Beacon shoal	78	10
Sun's centre Sun's true altitude 30° 10′	} 84	26
Beacon point	107	0
⟵ Pier	325	0

→ Pier	337°	30'
← Cliff	348	0
Zero	360	0

At Signal hill △.

Zero, Bluff △	360°	0'
→ Bluff	3	30
Rock awash	23	0
Extreme of Beacon shoal	30	30
Beacon point	51	40
Beacon △	55	0
← Pier	76	50
→ Pier	90	0
→ Cliff	93	0
← Cliff	324	10
Bight of sand	327	50
Rock on beach	335	20
Extreme of sand spit	345	30
Zero	360	0
Angle of depression to rock awash	1	47
Height of theodolite	5 feet.	

At Bluff △.

Zero, Signal hill △	360°	0'
→ Cliff	16	0
Extreme of sand spit	20	40
Bight of sand	28	25
Rock on beach	57	10
Extreme of Beacon shoal	280	20
Beacon point	292	0
Beacon △	295	0
Rock awash	321	5
Direction of pier	327	20
← Cliff	344	0
Zero	360	0

XIV.]   METHOD OF PLOTTING A SURVEY.   285

From Beacon point to inner end of pier and on to Cliff the coast is steep-to, the Cliff is of moderate height: from east end of cliff to the Rock on beach it is sand, and from thence gradually rises to a high cliff at the Bluff.

Beacon shoal is one cable in breadth, with patches of less than 6 feet at low water ordinary spring tides.

Sketch in the coast line; draw the true and magnetic meridians, and give the height of the Signal hill △.

Scale, 1 mile = 6073 feet = 4·37 inches.

Results. Height of Signal hill 155 feet.
Variation 10° W.
Signal △ to outer edge of pier = 3·03 inches.
Beacon △ to rock awash   = 2·97   ,,
West end of cliff to sand spit = 3·95   ,,

No. 17.   *Final,* 1880.

In survey of a harbour, stations were made on *Cliff* at north entrance point, *Hill* on west side, and *House* on south entrance point. *Cliff* △ bore N. 20° 30′ E. (true) from *Hill* △ and was found by calculation from a measured base to be 18,020 feet distant.

At *Hill* △.

Zero *Cliff* △	360° 00′
Rock awash	14 40
Islet (10 feet high) depressed 0° 55′	36 20
Point, south side of harbour	51 00
House △	64 20
Windmill φ ←— beach	327 50
Church	336 20
—→ beach	347 10

At Cliff △.

Zero, Hill △	360° 00'
Church	38 10
Windmill	41 10
Islet	284 00
House △	299 50
Rock awash ϕ Point, south side of harbour	336 00

At House △.

Zero Hill △	360 00
Point, south side of harbour	9 10
←—— beach	14 00
Windmill	31 20
Church	34 30
——→ beach ϕ Rock awash	38 20
Cliff △	55 30
Islet	76 30

Between Cliff △ and north extreme of beach the coast, composed of cliffs, forms two small bights; the beach is sandy and nearly straight. Hill △ is half a mile inland, and stands on summit of a hill which slopes gently to northward, but falls steeply to southward. Point, south side of harbour has several detached rocks which do not cover lying off it, and the shore forms a deep bight on either side.

Protract on scale of one inch = a mile of 6,080 feet; find height of Hill △ and sketch in details.

Results. Distance of Islet from Hill = 19,304 feet.
Height of Hill = 319 feet.
Windmill to Islet = 3·11 inches.
Rock awash to Point = 1·18 inches.

No. 18. *December,* 1880.

In a running survey of an island the ship steamed on the following courses, angles being taken at the position started from, and at the termination of each course :—

Mag. course.	Dist.
N. 75° E.	1·75 miles.
North	1·5 ,,
N. 60° W.	1·35 ,,
S. 43° W.	2·75 ,,

At Ship's 1st position.
Peak N. 12° E.
S.W. point 30° 0' Peak.
Peak 43° 0' S.E. point.

At Ship's 2nd position.
Peak N. 60° W.
⎯ island 23° 0' Peak ϕ S.E. point.
Peak 24° 0' Hill.
Do. 30° 0' North point.

At Ship's 3rd position.
Hill S. 88° W.
S.E. point 39° 0' Peak.
Peak 22° 30' Hill
Do. 40° 0' North point.

At Ship's 4th position.
Hill S. 10° E.
Peak 18° 0' S.W. point.

At Ship's 5th position.
North point 6° 0' Hill.
S.W. point ϕ Peak.

Elevation of peak 1° 34' 20" on the arc.
,,   ,,   1° 38' 20" off ,,

288   MARINE SURVEYING.   [CHAP.

Variation 25° W.   Scale 2 inches = 1 mile = 6,060 feet.

Courses and bearings are magnetic.

Between the S.W. and S.E. points the coast curves to the southward, and is a moderately high cliff; from the S.W. and S.E. points the coast trends directly towards the north point, the West coast is sand, and the East coast shingle.

Sketch in the island and give the true bearing of the peak from the 5th position, and its height.

N.B.—*Place the ship's 1st position 4½ inches from the right and 3 inches from the bottom of the paper.*

Results.   Peak bears N.E. (true) from ship's 5th position, is
   1·77 miles distant, and is 301 feet high.
   Island is 1·55 miles from N. to S.W. point.
   „   1·50   „   N. to S.E.   „
   „   1·50   „   S.W. to S.E.   „

No. 19.   *Final,* 1881.

In survey of a group of islands, a theodolite station was made on *High island, Ship* anchoring eastward of the group, and the *Cutter* anchoring northwest of them.

At *High island* △.

Zero, Distant peak (bearing N. 57° W. true) 360° 00'.
Rock awash	26° 20'
Cutter	35 00
←— Flat island	65 30
—→ „	79 00
⎰Ship's mainmast at water-line	145 40
⎱ „ depressed	0 48 10"
Breaker	213 00
←— Red rock	235 00
—→ „	239 00

XIV.]   METHOD OF PLOTTING A SURVEY.   289

At *Ship* (sextant station).

⟵ High island	9° 30′	High island △	10° 30′	⟶ High island.
Breaker	26 50	,,	15 00	Rock awash.
⟶ Red rock	33 50	,,	23 30	⟵ Flat island.
⟵ ,,	37 00	,,	32 30	Cutter (seen over Flat island).
		,,	34 30	⟶ Flat island.

At *Cutter* (sextant station).

⟵ High island   10° 30′   High island △   6° 00′   Rock awash.
⟶ Flat island   23 00   ,,   11 00   ⟶ High isl.
Ship (seen over
  Flat island)   36 50   ,,

Horizontal distance between High island △ and ship's mainmast at waterline, 9,010 feet.

High island is cliffy and circular in shape with a conical summit.

Protract the above on a scale of 2 inches = a nautical mile of 6,080 feet, and find height of High island, drawing scale and subdividing it into cables.

Note.—*High island station should be placed about the centre of the paper.*

Results. Height of High island 126 feet.
          Cutter is distant from High island 2·69 inches.
          Breaker ,, Ship   2·73 ,,
          Red rock from High island   2·20 ,,

MISCELLANEOUS EXAMPLES AND QUESTIONS.

(1) Explain the different methods employed in ascertaining the length of a base line.   (B_1 and B_2, June, 1881.)

(2) Draw a scale of longitude corresponding to latitude 58° 20′ N., one mile of latitude being represented by 2·2 inches.
        Result. Scale required is 1·16 inches.   (Do.)

(3) During a voyage from La Guayra to Cartagena, calling at Puerto Cabello and Curaçoa, the following observations were made for errors of chronometer and rates:—

  At La Guayra on May 22, chron. fast $4^h\ 33^m\ 7^s\cdot8$
    ,,   ,, 28, ,,   4 33 02·2
  Puerto Cabello on June 5, ,,   4 37 19·6
  Curaçoa   on June 12, ,,   4 41 00·1
  Cartagena on June 25, ,,   5   7 . 22·2
    ,,   ,, 29, ,,   5   7 18·1

Required the Meridian Distances of the three latter places from La Guayra.

    $0^h\ 4^m\ 25^s\cdot16$; $0^h\ 8^m\ 12^s\cdot45$; $0^h\ 34^m\ 47^s\cdot25$. (1881.)

(4) From Flagstaff a Tree bore N. 75° W. 5,500 yards, and a Church N. 84° E. 6,100 yards.

' The following angles were taken from a ship to fix her position :—
    Tree 50° Flagstaff 45° Church.

Required the bearing and distance of the flagstaff from the ship.
    Scale ½ inch = 1,000 yards.

Project this question by the one-circle, and the two-circle methods.   (B_1 and B_2, Dec. 1880.)

(5) From a Beacon, Green point bore S.W. by S. 5,830 yards, and a Cliff S.E. by S. 6,270 yards.

The following angles were taken to fix the position of a shoal:—
    Green point 120° Beacon 92° Cliff.

Required the bearing and distance of the shoal from Green point.
    Scale ½ inch = 1,000 yards.

The two-circle, and the straight line methods to be applied.
                  (Do.)

(6) Explain the different methods used in marine surveying for determining the heights of hills. (B_1 and B_2, Final, 1880.)

(7) Distinguish between the *Theoretical* and *Accidental* Errors to which Observations are liable, and give examples of each kind.

(8) Draw the symbols employed on Admiralty charts to represent,—
 1st, A flood stream setting 2 knots at springs;
 2nd, A rock awash at low water.
(B_1 and B_2, Final, 1880.)

(9) The following angles were taken to fix the position of a Shoal:—
 West Point 115° North Patch 88° East Point.

The distance between West Point and East Point was 7,400 yards; from West Point to North Patch 6,180 yards; and East Point to North Patch 6,500 yards.

Required the distance of the Shoal from West Point.

Scale ½ inch = 1,000 yards.

N.B.—Project this question by the straight line method.
(Do. Dec. 1877.)

(10) Explain the different methods used for measuring a Base Line in marine surveying. (Do. Final, 1878.)

(11) Explain the construction of a *Current-Log*, and state the Rule adopted to find the rate of a Current.

(12) From summit A, 239 feet above the sea surface, B was elevated 5° 45' and C depressed 2° 12'.

Horizontal distance between A and B 1,892 feet.
  „ „ „ A and C 1,530 feet.

Required the heights of B and C above the sea surface.

B 429 feet, C 180 feet.
(Do. Final, 1878.)

(13) From Gunboat in 3½ fathoms, Frigate in 6 fathoms bore S.S.W. (true), distant 2 miles, and Sloop in 5 fathoms bore W. ½ S. (true), distant 1·5 miles.

At Launch in 2½ fathoms.
Gunboat 24° 30' Sloop 35° 18' Frigate.

Soundings in fathoms at equal distances between Sloop and Launch $4\frac{1}{2}$, 4, 3. Reduction to low water 9 feet.

Protract on scale of 2 ins. = a mile, and place position of Launch and the *reduced* soundings on the plan. (Do.)

(14) From a Beacon, West Point bore S.W. 6,340 yards, and East Cape S.S.E. $\frac{1}{2}$ E. 5,925 yards.

The following angles were taken to fix the position of a ship:

West point 122° Beacon 128° East Cape.

Required the bearing and distance of the ship from the Beacon.

Scale $\frac{1}{2}$ inch = 1,000 yards.

(Do. April, 1879.)

(15) From Point △, Cairn bears North (true) 3,040 feet, and Bluff bears S. 22° E., 2,810 feet; project these positions on a scale of 4 inches = one mile = 6,080 feet, also the following line of soundings, reduction to low water 3 feet :—

						Fms.
$4^h\ 0^m$	Cairn	44° 30′	Point △	32° 0′	Bluff	7
		7	$6\frac{3}{4}$	$6\frac{1}{2}$		
$4^h\ 12^m$	do.	63° 0′	do.	44° 30′	do.	$6\frac{1}{2}$.
						s

(Do. Final, 1879.)

(16) Explain how a Base Line should be measured by Sound, and the circumstances under which this method may be found useful. (Do. Final, 1879.)

(17) If m = height of a masthead, a_1 = depression of the masthead, a_2 = depression of the waterline, and h = height of the position of observation above the sea, prove the formula

$$h = m\{1 + \cos a_2 . \operatorname{cosec}(a_2 - a_1) . \sin a_1\}.$$

If the position of observation is *below* the masthead, and a_1 = angle of elevation of the masthead, prove that

$$h = m\{1 - \cos a_2 . \operatorname{cosec}(a_2 + a_1) . \sin a_1\}.$$

XIV.]   METHOD OF PLOTTING A SURVEY.   293

(18) Forts A, B, C are equidistant from each other 4 miles. A to the northward bears N. 40° W. (*mag.*) from B.

   At c.   At d.
 C 122° 50′ A 93° 45′ B. C 80° 30′ A 82° 0′ B.
     Variation 19° Easterly.

Protract on a scale of 1·2 ins. = a mile, and mark the positions of c and d.   (Do. Final, 1879.)

(19) Breadth of river Thames at Greenwich on a true north bearing from landing stage = 350 yards. Soundings *in feet* taken on true north bearing, 50 yards apart, as follows:—

    2, 7, 11, 17, 16, 14, 5, 2.

Reduction to L.W. average spring tides 2 feet. Vertical scale 1 inch = 10 feet. Horizontal scale 1 inch = 100 yards.

Draw the section of the river for low water at average spring tides.   (Do.)

(20) From Rock point, Green point bore West 5·65 miles, and a Peak N. by E. 4·755 inches.

The following angles were taken from a Ship to fix her position:—

    Green point 70° Peak.
    Do. 60° Rock point.

Required the bearing and distance of Rock point from the Ship.
    Scale ½ inch = 1 mile.

N.B.—This example is to be projected by the two-circle method.
         (Do. Dec. 1877.)

(21) Explain with the aid of a diagram (scale 1 inch to 4 feet) the terms, Spring rise, Neap rise, and Neap range; Spring rise is 15 feet and the Neap rise proportionate. (Do. Final, 1876.)

(22) Explain the meaning of "Surveying by means of the Back Angle."

(23) At X, in the same line with, and equidistant from stations A and B, the height by level on *back* staff at A read 2·5 feet, that on *forward* staff at B 5·6 feet; in the same line, at Y, equidistant

from B and C, the height on *back* staff at B read 4·5 feet, that on *forward* staff at C 5·3 feet. Required the difference of height of the ground at A and C.

Result. C is 3·9 feet below A. (Do. Final, 1876).

(24) The position of a Shoal was fixed by the following angles :—

Flagstaff 75° Tree 120° Beacon.

From the Tree, Flagstaff bore S.S.W. $\frac{1}{2}$ W. 5·87 miles, and Beacon S.E. $\frac{1}{2}$ S. 8·13 miles.

Required the bearing and distance of the Shoal from the Tree.

Scale $\frac{1}{2}$ inch = 1 mile.

The straight line method of projection is to be here used.

(Do.)

(25) From a hill, White rock bore S.W. 5,680 yards, and Green patch S.E. by E. 7,140 yards.

The following angles were taken from a ship to fix her position :—

White rock 99° Hill 112° Green patch.

Required the bearing and distance of the hill from the ship.

Scale $\frac{1}{2}$ inch = 1,000 yards.

The two-circle method of projection is to be used. (Do.)

(26) From a beacon a spire bore N. 85° W. 4,550 yards; and a tower N. 10° E. 5,210 yards.

The following angles were taken to fix the position of a Shoal :—

Spire 56° Tower.
Spire 42° Beacon.

Required the bearing and distance of the Shoal from the Beacon.

Project this question by either method.

Scale $\frac{1}{2}$ inch = 1,000 yards.

(Do. Dec. 1879.)

(27) From Flagstaff, Windmill bore N. 33° E. (true), 1·1 miles, and Church S. 34° 30′ E., 1·44 miles.

Project these positions on a scale of 2 inches to a mile; also the following line of soundings, the reduction to low water being 2 feet:—

$3^h\ 40^m$ Windmill 58° 30′   Church $6\frac{1}{2}$ fms.
        Flagstaff 56 10   do. m.
        $6\frac{1}{4} - 6\frac{1}{4} - 5\frac{3}{4}$
3 45 Windmill 74 50   Church $5\frac{1}{2}$
        Flagstaff 72 30   do. s.m.

(Do. Final, 1875.)

(28) From a Hut, West Point bore S. 20° W. 6,200 yards; and East Point S. 50° E. 6,500 yards.

The following angles were taken to fix the position of a Rock:—

West Point 85° Hut 118° East Point.

Required the bearing and distance of the Rock from the Hut.

Project the above by the straight line and the two-circle methods.

Scale ½ inch = 1,000 yards.

(Do. Dec. 1879.)

(29) From a Cairn a Flagstaff bore N. 75° W. 6,185 yards, and White patch N. 84° E. 5,850 yards.

The following angles were taken to fix the position of a Ship:—

Flagstaff 41° Cairn 52° White patch.

Required the bearing and distance of the Ship from the Cairn.

Project the above by the one and the two-circle methods.

Scale ½ inch = 1,000 yards.

(Do. Dec. 1879.)

(30) From South Bluff, West Point bore W.N.W. 4,890 yards, and East Point E.N.E. 4,650 yards. The following angles were taken to fix the position of a Wreck:—

West Point 62° South Bluff 39° East Point.

Required the bearing and distance of the Wreck from the South Bluff.

Scale ½ inch = 1,000 yards.

The one and the two-circle methods of projection to be applied.

(Do. Dec. 1876.)

(31) From A in latitude 15° 10′ 30″ N., longitude 80° 12′ 0″ E., B bears N. 17° E. (true) 22,560 feet; one mile of latitude = 6,050 feet. Required the latitude and longitude of B.

Results. Lat. of B 15° 14′ 5″ N.
Long. of B 80° 13′ 7″ E.

(Do. Final, 1877.)

(32) From a hill, the angle of depression to the summit of an islet, 30 feet above the sea and distant 2,500 feet, is 2° 11′; height of eye 5 feet. Required the height of the hill.

Result. 120 feet. (Do.)

(33) Define the terms, tidal wave and tidal stream. (Do.)

(34) Required the natural scale of a plan, drawn on a scale of 6 inches = one mile = 6,082 feet. *Result.* $=\frac{1}{12164}$.

(Do.)

(35) From Sand hill, Wedge bears N. 39° W. (true) 4,550 feet, and Bluff bears N. 78° E., 5,990 feet. Project these positions on a scale of 4 inches = one mile = 6,070 feet; also the following line of soundings, reduction to low water 9 feet :—

3ʰ 30ᵐ Wedge 33° 20′ Sand hill 41° 40′ Bluff 10.
$9\tfrac{3}{4} - 9\tfrac{1}{2} - 9\tfrac{1}{2}$
3 42 do. 44 20 do. 39 40 do. 9¼.
s

(Do.)

(36) Investigate a Formula for computing the *Horizontal Angle* between the sun and a given object, the latter not being in the horizon.

(37) Given two positions on a Mercator's chart of the world shew how to find the distance between them.

(38) Does a straight line on a Mercator's chart of the world ever represent the shortest distance between two positions?

(39) Given the cross-bearings and deviation of the compass, place the ship's position on a chart.

(40) Given a bearing of an object and an angle between it and another object, lay down the ship's position.

(41) Lay off a course to pass clear of a known danger, allowing for variation and deviation.

(42) What observations would you make for the Latitude, Longitude, and Variation of the Compass, for a given place?

(43) Explain the term "tangent" as used in surveying.

(44) What do you understand by the term "angle of danger"?

(45) Lay down a position by a "circle and line of direction," knowing the angles between three points.

(46) Lay down a position by circles only, knowing the angles between three points.

(47) Define an "ill-conditioned" triangle.

(48) State the difficulty sometimes experienced of "being on the circle."

(49) Mention the method of getting out of this difficulty.

(50) What precaution must be taken when observing a small angle with a sextant?

(51) Draw a cluster of rocks, some above water, others awash, with encircling danger line, extending about a mile east and west, and half a mile north and south. Scale 2 inches = a mile.
(Dec. 1878.)

(52) Draw the symbols used on Admiralty charts to represent a Church, a Windmill, and an Observation spot. (June, 1880.)

(53) Draw the symbols to denote an *Anchorage for large vessels*, and an *Ebb-tidal stream*.

(54) Draw the symbols used to represent:
  (i) A *can buoy* painted red and white in horizontal stripes.
  (ii) A *rock awash at low water*.
  (iii) The position of a *lighthouse*. (Oct. 1880.)

(55) Write the abbreviations for *sand* and *broken shells* in describing the nature of the bottom. (Dec. 1880.)

(56) Draw a coral reef of circular shape, half a mile in diameter, with a lagoon in the centre, and rocks awash extending half a mile from its northern side. Scale 2 in = a mile.
(April, 1881.)

(57) Draw the symbols used to represent a *coral reef*, a *marsh*, and *an anchorage for small vessels*. (June, 1881.)

(58) Describe any methods you are acquainted with for finding the rate and direction of a *surface current* from a ship at anchor; also of an *under current*. (Feb. 1879.)

(59) What steps are necessary in order to determine with accuracy, by Chronometer, the Meridian Distance between two places? (Dec. 1877.)

(60) Describe the different methods of measuring a Base, with the necessary precautions for accuracy in each case.
(Feb. 1875.)

(61) Does a Nautical Mile vary in length in different parts of the earth? (Sept. 1877.)

(62) Describe the use of the Level and Levelling-Staves as applied to nautical surveying. (May, 1879.)

(63) Explain and illustrate what is meant by a *landmark*, *leading mark, clearing mark, danger angle*, and *transit of land objects*. (Dec. 1879.)

(64) Does the direction of the tidal stream always change at the time of high water? (April, 1880.)

(65) Explain the abbreviations H.W.F. and C.; at which seasons do the highest tides occur? (Sept. 1880.)

(66) Draw a compass as shewn on Admiralty charts; diameter 3 inches, variation 26° W. (May, 1881.)

(67) Define the term *Neap range of tide*. Is the level of the water higher or lower at low water neaps than at low water springs? (Aug. 1880.)

(68) How is the True Bearing of a point of land determined by means of *sextant* observations? (Aug. 1881.)

(69) Draw a compass similar to those usually shewn on Admiralty charts, subdividing one quadrant. Variation 23° 20′ E. (Aug. 1881.)

(70) State how a Theodolite may be used in swinging ship for deviation. (Aug. 1876.)

(71) What do you understand by the Index Error of a Theodolite?

(72) What is a *Day Mark*?

(73) Distinguish between *Absolute* and *Dependent* Heights.

(74) Distinguish between the terms *regular* and *irregular* plotting.

The following may be regarded as typical questions in the case of a Vivâ Voce *Examination in the Subject.*

(75) Mention the best means of finding the Error of Chronometer at sea.

(76) If the A.M. sights of Equal Altitudes are lost, are the P.M. sights completely lost also? What is the disadvantage attending the loss of the Forenoon sights?

(77) Mention, in their proper order, the Adjustments of the Sextant.

(78) What is the use of the Graduations in a Compass?

(79) Mention the *Advantages* and *Disadvantages* of this Instrument when compared with a Theodolite.

(80) Explain how you would set up a Theodolite to take an "arc."

(81) Lay off on a Chart an Observed Bearing, using Parallel-rulers.

(82) Note the best Headlands in a given Harbour for fixing a position by the Three-point Problem.

(83) What substitute would you suggest in lieu of a Station Pointer?

(84) Read off and explain the following note about a Light on the Coast:

120 ft. Rev. 1 min. vis. 16 m.

(85) Can a Light be seen from a point *on* the Circle described on the Chart to shew the extent of illumination?

(86) Distinguish between a Plan and a Chart; how would you know one from another if placed in your hands?

(87) Mention the respective advantages of Mercurial and Aneroid Barometers.

(88) What is the Datum Line to which Soundings are to be reduced?

(89) Is there *any* difference between the several Compasses engraved on a Chart of considerable extent?

INDEX.

The Numbers refer to the Paragraphs.

Abbreviations, 2
Absolute Height, 160
Admiralty Tide Tables, 198
Afternoon Sights in Equal Altitudes, 253
Age of a Tide, 190
Airy's Method of Compensation, 231
Alternating Light, 7 (7)
Aneroid, 159
Angle, Methods of protracting, 44
— by Scale of Chords on Protractor, 33
— by "Versine Method," 45
— to repeat, with a Theodolite, 86
— to observe a vertical, with ditto, 87
— Horizontal, 128
"Annual Trials," 232; 233
Anti-Lunar Tide, 167
Anti-Solar Tide, 167
Aphelion, 178
Apogee, 178
"Arc," to observe, with a Theodolite, 85
Assumed Base, 139

Bach's Base Measuring Apparatus, 111

Balance of a Chronometer, 231
Bar, length of, involves three elements, 107
Barometer, Correction of, 93
— Contracted Scale of, 96
— Description of, 93
— Fortin's, 95
— Height by, 159
— Marine, 97
— Method of Suspending, 98
— Mountain, 157
— Reading off, 100
— Packing for Carriage, 101
— Vernier of, 99
— Effects of, on Tides, 182
Base, Assumed length of, 139
— Direction of, 127
— Elements of, 117
— Examples of accurate, 104
— Method of Increasing, 133
— Length, by Astronomical Observations, 125
— — by Chaining, 126
— — by Masthead Angle, 122
— — by Patent Log, 124
— — by Velocity of Sound, 123
— Method of Measuring, 126

MARINE SURVEYING.

Base, Madrid, 115
— Names of, in Ordnance Survey, 114
— Nature of the ground for, 120
— Ratio of length to Size of Survey, 113; 119
— Reduced to the Sea level, 116
— Use of, 102
— of Verification, 104
Bench Mark, 13; 145
Bernouilli's Theory of Tides, 164
Borda's Method, 108
Bottom, Symbols to denote nature of, 1
Buoy and Nipper, 226
Buoys, Colouring, 4
— Names, Shapes, and Colours of, 3
— Regulations for use of, 5

Capacity, Correction for, 156
Capillarity, 156
Catodioptric System, 8
Catoptric System, 8
Centre, Reduction to, 141
Chain, Gunter's, 76
— Surveying, 77
Change Tide, 191
Chart, Explanation of, 58
— Mercator's, 64
— Method of Constructing, 68 ff.
— Official Number of, 12
— Plane, 65
Check Level, 145
Chords, Angle projected by, 33; 45
— Line of, 36
— Scale of, 32
— Use of ditto, 33
Chronometer, Balance wheel, 231
— Comparison of, 242; 243
— Compared with Sidereal Clock, 246

Chronometer, compared with Deck Watch, 244
— Effect of iron, 238, *note*
— Faulty, how detected, 244
— History of, 228
— Inside of, 230
— Journal, 244
— Life of, in Royal Navy, 235
— Method of Winding, 241
— — Setting going, 241
— Ought to be wound regularly, 240
— Packing of, for transport, 236
— Rating of, 242
— Room, in Royal observatory, 233
— Sources of Error, 239
— Stowing on board, 238
— Selection of, for Royal Navy, 234
— Temperature, effect of, 240
— Transporting from Shore to Ship, 237
Chronometrical Thermometer, 233
Circle, on the, 53
Clearing Mark, 12
Colby's Apparatus, 109
— Method of using, 110
Collimation, To test, 80
— To correct, 80
Coloured Sectors, 8
Compensation Apparatus, 109
— Microscopes, 110
Conjunction, 178
Constants, Tidal, 195
Contours, 161; 162
Conventional Signs, 10
Convergency of Meridians, 63

INDEX. 303

Corrected Establishment of the Port, 191
Cotidal Lines, 172; 173
Course, on a Chart, 73
Current Log, 220
— Surface, 220
— Tidal, 175
— Symbol for, 12
— Under, 221; 222
— of the Flood, 177

Danger Angle, 56
— Hidden, 218; 219
Datum Line, Def., 145
— in Soundings, 213
— in Ordnance Survey, 145
Day Mark, 12
Dependent Height, 160
Diagonal Scales, 19
Dioptric System, 8
Direction of the Base, 127
Discharge of a River, 223
Distance on a Chart, 73
Division of a line into parts, 43
Diurnal Inequality, in Heights, 180
— in Times, 181
Double Half Day Tides, 187
Drummond Light, 138
Dynamical Theory of Tides, 164

Ebb of a Tide, 176
Elements of a Base, 117
Error, Secondary, of a Chronometer, 232
Establishment of the Port, Def., 190; 191
— — Method of finding, 192
— — Corrected, 191
— — Vulgar, 191
Equal Altitude Observations, 252; 253
Equilibrium Theory of Tides, 164
Evolution, Tidal, 205

First and Second High Waters, 194

Fixed Light, 7 (1)
Fixed and Flashing Light, 7 (3)
Flashing Light, 7 (2)
Flow of a Tide, 176
Fortin's Barometer, 95

Gauge, Tide, 201
Gravel Beach, 10
Ground Log, 226
Gunter's Chain, 76

Hachure Lines, 163
Harrison's Watches, 229
Heat Test of Chronometers, 233
Height, by Aneroid, 159
— by Barometer, 159
— by Spirit-Level, 147; 148
— by Theodolite, 153
— by Thermometer, 160
— Formulæ for, by Barometer, 159
— of a Tide, 183; 184
— Absolute, 160
— Dependent, 160
Hidden Danger, 218; 219
High Water, Def., 165; 177 *note*
— First and Second, 194
— Time of, 196; 203; 204
Horizontal Angle, 128
Hutton's Rule, 158

Illuminating Apparatus, 8
Increasing the Base, 133
Indeterminate Case, 53
Index Error, in Sextant, 80
— Theodolite, 84
Inequality, Diurnal, 180; 181
Inferior Tide, 167
Intermittent Light, 7 (5)

Journal, Chronometer, 244

Lagging of a Tide, 186
Land Base, 126

Landmark, 12
Laplace's Theory of the Tides, 164
Lateral Distance, 34
Latitude on a Chart, 73
Law of the Heights of High Water, 193
Lead Lines, 212
Leading Mark, 12
Least Reading by Vernier, 23 ff.
Legendre's Theorem, 140, *note*
Length of Base, 119 ff.
Level, Apparent, 144
— True, 144
— Difference of, 144
— Check, 145
— Spirit, 88; 148
— Adjustments of, 88
Levelling, Def., 144
— by Aneroid, 159
— by Barometer, 155
— by Theodolite, 153
— by Thermometer, 160
— Simple, 146
— Compound, 146
— Staff, 89
— Precautions in holding ditto, 89
Light, Drummond, 138
— Alternating, 7
— Fixed, 7
— Fixed and Flashing, 7
— Flashing, 7
— Intermittent, 7
— Occulting, 7
— Revolving, 7
Light Vessels, 9
Lights on Coasts, 6; 7
Limb of a Sextant, to ascertain if correctly graduated, 29
Line of Chords, 36
— Lines, 35
— Polygons, 37
— to divide a, 43
Lines of position, 55

Lines, Cotidal, 172; 173
Log, Current, 220
Log, Ground, 226
Longitude on a Chart, 73
Low Water, Def., 165; 177, *note*; 203
Lunar Day, 169
Lunation, 178
Lunitidal Interval, 191

Madrid Base, 115
Map, Def., 58
Marquois Scales, 38
— use of, 39
Masthead Angle, 122
Mean level of Sea, 183
Mercator's Chart, Advantages and Disadvantages of, 64
— — Construction of, 68 ff.
— — Explanation of, 64
Meridian, Primary and Secondary, 248
— — use of, 249
— Distance, Def., 247
— — by Telegraph, 251
— — by Chronometers, 252
— — Examples of, 256
Middle Latitude, Scale of Chart for, 75
Mile, Nautical, Def., 66
Mountain Barometer, 157
— how delineated on a Chart, 163

Natural Scale, 17
Nautical Mile, 66
— varies in length, 66
Neap Tides, 165; 170
Nodal Line, 174

Observation Spot, 2
Observations with Sextant, 80
— Tidal, 200
Observatory Station, 13

INDEX.

Observatory, Latitude and Longitude of, 118
Occulting Light, 7 (6)
Octant Tides, 189
Official Number of a Chart, 12
Opposition, 178
Ordnance Survey, Scales used in the plans of, 18

Parallelism of a River, 217
Perigee, 178
Perihelion, 178
Periods of a Tide, 12
Perpendicular, to draw a, 41
Personal Equation, 131
"Piling-up" triangles, 133
Plan, Def., 58
Plane Chart, 65
Plotting, Method of, 260
Pole, Ten-feet, 90
Polygons, Line of, 37
Position, fixed by straight line and one angle method, 50
— by straight line and one circle method, 52; 55
— by straight line method, 52
— without drawing circles, 55
— Lines of, 55
Primary Meridian, 248
Priming of a Tide, 186
Primitive, 59
Problems, Useful, 57
Projection of a Point, 59
— Artificial and Natural, 59
— Central, 62
— Mercator's, 64
— Orthographic, 60
— Stereographic, 61
— by Chords, 44; 45
Protracting Scales, 30; 31
Protractor, Scales on, 15; 30

Quadrature, 178

R. M. S.

Range of a Tide, 183; 185
Reduction to the Centre, 141
— of Soundings, 213
Retard of a Tide, 190
Revolving Light, 7 (4; 6)
Rhombus Method of finding a distance, 57
Rip, Tide, 12
Rise of a Tide, 183; 184
River, Discharge of, 223
— Elevation along a tidal, 217
— Sectional area of, 224
— Velocity of, 225
Ruling-pen, Precautions in using, 72

Sandy Beach, 10
Scale, Def., 14
— of Chords, 32; 33
— Diagonal, 19
— of Equal parts, 15
— — use of, 16
— — construction of, 17
— Marquois, 38 ff.
— Natural, 17
— Plain, 15
— Protracting, 30; 31
— of Rhumbs, 31; 33
— of Secants, 31
— for a plan, how determined, 90
— Contracted, in Barometer, 96
Sea Base, 126
Sea, Mean level of, 183
Secondary Meridian, 248
— Error of Chronometer, 232
Section, Def., 144
— to lay down a, 149; 154
— Examples of, 150
Sectional area of a River, 224
Sector, 34 ff.
Sectoral Lines, 35
Segment of a Circle, to describe a, 48
Semi-lunation, 178
Semi-menstrual inequality, 191
Sextant, Adjustments of, 79; 80

20

Sextant, Limb of, incorrectly graduated, 29
Shingle Beach, 10
Single Day Tides, 187
Slack Water, 176
Sound, Velocity of, 123
Soundings, object of, 206
— to fix, 208; 211
— must be reduced, 213
— formula for reducing, 215
— in a river, 216; 217
— Symbols for Lines of, 11
Sounding Pole, 207
Spherical Excess, 140
Spirit Level, 88
— Adjustments of, 88
Spring Tides, 165; 170
Staff, Levelling, 89
Stand of a Tide, 176; 177 *note*
Standards, Geodetic, 105
— "à traits," and "à bouts," 105
— in Ordnance Survey, 105
Station Pointer, 91
— — Method of using, 92
Straight Line Method, 55
Straight Line and One Angle Method, 50
Stream, Tidal, 175
Struve's Method, 108
Superior Tide, 167
Surface Current, 220
Survey, Points to be observed in making a, 259
Surveying Chain, 77
Suspending a Barometer, 98
Sweeping for a hidden danger, 219
Symbols, Chap. I. *passim*

Tachometer, 225
Temperature, to evade effects of, on a Measuring Bar, 108
Ten-feet Pole, 90
— Scale for, 90

Theodolite, Description of, 81
— Adjustments of, 84
— Motions of, 82
— Parts of, 81
— Method of using, 85
— To repeat an Angle with, 86
— To observe a vertical Angle by, 87
— Levelling by, 153
Thermometers in Ordnance Survey, how Compared with the Standard, 106
— Height by, 106
Three-point Problem, 51
— — Six cases of, 52
— — Selection of the objects, 53
— — Best position of the "Points," 54
Tidal Day, 186
— Constants, 195
— Current, 175
— Stream, 175
— Observations, 200
— Evolution, 205
— Table, Elements of, 200
— Wave, 175
Tide, Flow and Ebb of, 165; 176
— Superior and Inferior, 167
— Spring and Neap, 165; 170
— Head, or End of, 174
— Priming and Lagging of, 186
— Age of, 190
— Retard of, 190
— Rise and Range of, 183; 184
— Octant, 189
— and Half Tide, 177
— and Quarter Tide, 177
— and Half Quarter Tide, 177
— Rip, 12
— Gauge, 201
Tides, Theories of, 164
— Caused by difference of attrac-

tion on opposite sides of the Earth, 166
Tides, when Greatest and Least, 179
— Single-day, 187
— Double Half Day, 187
— Effects of Sun and Moon in raising, 167 *note*
Time of High Water, how to find, 196
Topographical details in a Survey, 259
Transverse Distance, 34
"Trial Numbers," 234
Triangle, ill-conditioned, 132
— Well-conditioned, 132
Triangulation, Primary, 133
— Secondary, 133
— of Britain connected with that of the Continent, 142
— of Spain connected with that of North Africa, 143

Under-Current, 221, 222
Under-lined figures on a bank, 12

United States Coast Survey Apparatus, 111
Useful Problems, 57

Velocity of a River, 225
— of Sound, 123
— of a Tide, 12
— of a Surface Current, 220
— of an Under Current, 221; 222
Verification, Base of, 104
Vernier, 21 ff.
— of a Standard Barometer, 99
— Least Reading by, 23 ff.
— Exercises in, 25
Vertical Angle by Theodolite, 87
Vulgar Establishment of Port, 191
Vessels, Light, 9

Wave, Tidal, 175
Wind, effect of, on a Tide, 182
Winding a Chronometer, Method of, 241
Wreck, how Marked, 5

www.ingramcontent.com/pod-product-compliance
Lightning Source LLC
Chambersburg PA
CBHW030014240426
43672CB00007B/946